Principles of Highway Engineering and Traffic Analysis

Principles of Highway Engineering and Traffic Analysis

Editor

Shaithis Orlov

scitus academics

Principles of Highway Engineering and Traffic Analysis
Edited by **Shaithis Orlov**

Printed in 2017

ISBN: 978-1-68117-169-2
Library of Congress Control Number: 2015951145

SCITUS Academics LLC,
616, Corporate Way, Suite 2, 4766,
Valley Cottage, NY 10989

www.scitusacademics.com

Preface

The importance of highway transportation to the industrial and technological complex of the United States and other industrialized nations cannot be overstated. Virtually every aspect of modem economies, and the ways of life they support, can be tied directly or indirectly to highway transportation. From the movement of freight and people to the impact on residential, commercial, and industrial locations, highways have had, and continue to have, a profound effect on the world economy and societal development. In the United States, the manner in which highways have come to dominate the transportation system has been studied for decades as a cultural, political, and economic phenomenon. Without a doubt, the demand for unrestricted mobility and unlimited access to resources has played an important role and helped to quickly move highway transportation to its dominant position from the middle of the 20th century onward.

The construction of the interstate highway system remains to this day the largest infrastructure project in human history. At the time, it underscored the nation's commitment to the unrestricted mobility of its populace and to the economic opportunities that such a system would provide its industrial and service industries. Today, additional highway expansion and maintenance of existing highway systems continue to represent an enormous investment in public infrastructure an investment with an immeasurable impact on society in terms of mobility, economic opportunities, and environmental implications, including consumption of resources and pollution.

There is more demand than ever for highway engineers due to new highway projects throughout the country. This book interested engineers with the information needed to solve the highway-related problems that are most likely to be encountered in the field. It include road vehicle performance, the geometric alignment of highways, pavement design, traffic analysis, queuing theory, signalized intersections, the assessment of level of service, and traffic forecasting.

Table of Contents

INTRODUCTION

According to the World Health Organization, 600,000 people die and 15 million are injured every year in traffic accidents [1]. According to U.S. Department of Transportation's (USDOT's), motor vehicle crashes are the leading causes of death for people aged 1 to 33. Societal economic losses from these crashes are huge, estimated by the National Highway Traffic Safety Administration to exceed $230 billion in 2000 [2]. In Israel, the number of casualties from traffic accidents is higher than the total number of casualties in all Israeli wars [3].

Road accidents and their spatial dispersion, including counting and probability estimation models, have received increasing research attention in recent decades. This is due to their high cost and importance to society in terms of physical, mental and economic aspects. Many studies have examined various factors contributing to road accidents in attempt to formulate explanatory variables from different fields: *vehicle fleet* (distribution of fleet by category and age); *routes* (road length; distribution of lengths by road classes); *exposure* (kilometrage; kilometrage by road classes; kilometrage by road users; passengers' kilometrage by mode of transport; passengers' kilometrage by age and gender; traffic flow by road classes); *population* (population by age and gender; license holders by age and gender; percentage of novice drivers; density); *meteorology* (temperature; levels of precipitation; sunlight; ice); *traffic safety*(alcohol consumption; seat belt wearing rates; speed; helmet wearing rates; air bag equipment rates; speed limitations; minimum age for driving; safety interventions (safety policies); *economics* (household income; household final private consumption; consumption prices; wages; gross national product; unemployment; industrial production; active population; gas prices; gas consumption); *economics and traffic safety* (national expenditures in road engineering; national expenditures in road investment; national expenditures in road safety actions; national expenditures in road police); *miscellaneous* (costs of accidents; education level; criminal level; suicides; strikes) [1,4,5,6,7,8,9,10,11,12,13,14,15, 16].

Traffic volume, as measured by Annual Average Daily Traffic (AADT), is one of the most often cited exposure measures and explanatory variables. Researchers have studied it in depth in recent decades using a variety of methods. As AADT was supposed to have a linear effect, the first and most obvious approach used to study it, was that of the linear regression models. Standard models used least squares techniques, sometimes with a log-log or a quadratic formulation to achieve a better fit. The linear

CHAPTER 1

Geo-Based Statistical Models for Vulnerability Prediction of Highway Network Segments

Keren Pollak[1], Ammatzia Peled[1] and Shalom Hakkert[2]

[1]Department of Geography and Environmental Studies, University of Haifa, Mt. Carmel, Haifa 39105, Israel;
[2]Division of Transportation and Geo-Information Engineering, Technion-Israel Institute of Technology, Haifa 32000, Israel

ABSTRACT

This study describes four statistical models—Poisson; Negative Binomial; Zero-Inflated Poisson; and Zero-Inflated Negative Binomial—which were devised in order to examine traffic accidents and estimate the best probability estimating model in terms of future risk assessment at interurban road sections. The study was conducted on four sets of fixed-length sections of the road network: 500, 750, 1000, and 1500 m. The contribution of transportation and spatial parameters as predictors of road accident rates was evaluated for all four data sets separately. In addition, the Empirical Bayes method was applied. This method uses historical accidents information, allowing regression to the mean phenomenon so as to improve model results. The study was performed using Geographic Information System (GIS) software. Other analyses, such as statistical analyses combined with spatial parameters, interactions, and examination of other geographical areas, were also performed. The results showed that the short road sections data sets of 500 and 750 m yielded the most stable models. This allows focused treatment on short sections of the road network as a way to save resources (enforcement; education and information; finance) and potentially gain maximum benefit at minimum investment. It was found that the significant parameters affecting accident rates are: curvature of the road section; the region and traffic volume. An interaction between the region and traffic volume was also found.

formulation came under debate, with its interpretation and statistical assumptions being questioned. As an alternative method, the Poisson and Negative Binomial regression models were developed. These were considered more advanced due to the assumed Poisson distribution of road accident counts [17].

It is important to realize that a traditional application of the Poisson or negative binomial distribution alone does not address the possibility that more than one underlying process may be influencing crash frequencies. For instance, if the study segments are collected randomly, a preponderance of zero-crash observations will appear in the data because crashes are rare events. This over-representation of zero-crash observations in the data may erroneously suggest overdispersion in the data even though the Poisson distribution is actually otherwise correct [13]. Thus, the Zero Inflated Poisson regression model allows overdispersion in the data due to excess zeros when compared to the classic Poisson regression model [18].

Speed is one of the most investigated variable in road accidents studies. The author of [19] examined the effect of extreme speeds, road environment and geometry on traffic speed and accidents. During the research, prediction models for European Union roads were built. He stated, that much research has been done on this subject, but opinions still seem to differ among scholars as to whether the mean speed or speed variance affects accidents and if so to what extent. The authors of [15] raised it explicitly. They found that the accident rate rose in accordance with increasing speed on non-major roads. Yet, on main roads, it was found that the path width, the density of nodes and traffic, had higher correlations with the accident rate.

One of the greatest problems with the statistical modeling is the lack of variables linked directly with road infrastructure (except for national expenditures or investments in road engineering). Yet, it is commonly agreed that the design and maintenance of roads are not negligible in the accident process. Improvements and safety engineering are usually supposed to be included in the model trend without any explicit variable [1]. The authors of [17,18] were among the first who investigated the relationship between vehicle types and physical road characteristics using both linear and non-linear models. In their research, they examined the relationships between physical variables of the road (e.g., segment length, curvature, shoulder width and slope), transportation variables (e.g., traffic volume, truck travel), and the number of accidents involving trucks within a particular section. Upon testing the Poisson distribution models, their study revealed that when incorporated into very short road sections (<80 m), linear model accuracy was corrupted. Parameters such as traffic volume, curvature, and slope were found to have a positive correlation with traffic accidents involving trucks. Later on, few studies have combined physical parameters of the road in the model, such as [20].

The authors of [2,5,6,21,22] were among the first that incorporated the capabilities of Geographic Information System (GIS) with Road Safety Analysis. The author of [3] for example, developed a unique software package (Arc\Info-based) for road safety analysis. GIS offer advanced solutions for both area-wide and location-oriented investigations. It also enables making complex analyses and in-depth investigations relatively easy.

As mentioned above, previous studies did not study variables linked with road infrastructure, geometry and geospatial data, which can be extracted relatively in a simple way using Geographic Information Systems (GIS). In addition, the assumption was that if the models will produce significant results for the short road sections datasets, it will allow allocating resources in a focused and economical way, contrasting the way in which those are treated today. Therefore, the goal of the research was to develop a GIS-based prediction model for the assessment of traffic accidents in highway short segments. Using spatial and traffic-based parameters, the research focused on the following issues:

1. Lengths varying: can we acquire significant predictive models for short road segments (500 m). If so, target treatment on road sections, in terms of infrastructure and enforcement can be recommended.
2. Significant variables: Given the traffic and spatial data on a highway segment, what are the most significant variables affecting the rate of accidents for that segment?
3. Interactions between variables: Is there an interaction between the following variables: (a) Area and segment curvature: Does curvature have a different effect on the number of accidents in different areas? (b) Area and slope: Can we determine a different expected number of accidents for slopes in different areas? (c) Traffic volume and area (AADT): Is the effect of traffic volume on the rate of accidents inconsistent and dependent on a given area?

STUDY AREA AND METHODS

Study Area and Dataset Construction

The study was conducted on the main road network of Israel. The study uses traffic accident data, as reported by the Israeli Central Bureau of Statistics (CBS), for the years 2005 to 2007 [23]. The CBS database is based on information distributed monthly by the Israeli Police. These are only accidents with casualties. Other accidents, such as fender bender are not dealt with by the Israeli police. In the current study were integrated all accidents that occurred on highway's sections only (not on intersections). This left us with 2592 accidents in 2005; 2558 accidents in 2006 and 2373 accidents in 2007 (total of 7523 accidents). The accidents were positioned on the road network of Israel. This was done by an address matching process using road number and running length provided within each accident record. The actual address matching process was based first on fusing two GIS layers of roads and intersections (carrying the running length). Later, the actual address matching was applied by ArcGIS "make route event" tool. The accident running length is reported by the police examiners in a resolution of 100 m along the road.

Four initial data sets were generated from the original main road network of Israel. Each of these was built using different road section lengths: 500 m, 750 m, 1000 m, and 1500 m. After removing all segmented within urban regions and those adjacent the to the intersections, we were left with: 572, 1166, 1817 and 3259 segments at the "1500 m", "1000 m", "750 m" and "500 m" datasets, respectively. Each road section (at all four data set) was associated with characteristic values of spatial parameters, e.g., geographical area (north, center and south) and sub-region areas; slope; curvature; solar angle (sun glare) and with transportation characteristic e.g.,: number of accidents, AADT-Annual average daily traffic for weekdays: Sunday—Thursday.

Later on, for each data set, four statistical models were built using SAS software. These were: Poisson; Negative Binomial; Zero-Inflated Poisson; and Zero-Inflated Negative Binomial.

Modeling: Implementations and Considerations

Traditionally, building a predictive model for road accidents uses historical data for constructing the model. These predictions will then be compared with actual future observations of accidents (at specific, unchanged places). In the present study the database that was built is innovative and unique, incorporating dedicated parameters, as described in the previous section. As mentioned, one of those was traffic volume (AADT), which was found to be a significant parameter in many other studies in the field of road safety.

In the current study AADT was mostly updated manually according to CBS documentation (at a time consuming process). For each segment data for three years of study were updated. Sometimes, the CBS documents were lacking some information regarding segments AADT (no data for one year out of three research years or no data at all). Therefore, in order for not impairing the modeling quality while building the statistical modeling, only segments that had all needed data for the whole three research years (2005–2007) were integrate into the study. This resulted with instances in which constructing a probability estimating model using "historical" data of 2005–2006 and then compare the predictions to future observations (e.g., 2007 data) was not feasible in term of statistical model validation.

The offset variable used in poison regression models, takes count data (e.g., number of vehicles) and returns rates (count per unit), for example number of vehicles per km. At the current study, AADT was included into the prediction model as an independent variable but not an offset. This, as in the research we wanted to examine the relationship between AADT and the number of accidents. If the AADT was integrated as offset, it could not be done. Also the AADT was related to equal size road segments, so an additional normalization was not required.

The intention here was to ensure that data dispersion is maintained and that the variance is preserved. This was in order to verify that we modeled segments as "zero accidents" not only because the time period that was chosen was too short, rather than other reasons (such as the road section is actually safe). Therefore, each one of the four data sets (different length of segments) was randomly divided into two subsets. The first consisted of two-thirds of the events, and the second comprised the other one-third of the observations. The larger subset ("2/3") was used as a "training set" for generating the model, while the smaller subset ("1/3") was used as a "control set" in order to verify the results. An additional model was constructed from another random one-third of the data set to serve as the "1/3—full control model". This model incorporated all of the existing parameters in the data set and was used to examine significant results (regardless of the "training model"). All these were done in order to determine the models stability.

Section 2.4 will detail the "Cross validation" process that was performed using all three different models that were built (according to three groups mention above). These validations objective was to ensure, independently, which are the variables that are significantly contributing to car accidents. The aim of the validation and comparison of the different models (detailed at 2.4) was to ensure, that the "training prediction model" is not biased in its average prediction. Note, that this is not an indicator of the quality of the predictions. These will be required to be tested in future research.

In studies characterized by large data sets, examining the significance value is not enough. Thus, due to the impact of large quantity, sometimes even low-impact parameters appear to be significant. A solution for this problem is the use of Akaike Information Criterion (AIC) as a test for assessing the quality of the model. This criterion is based on the likelihood function, which compares the predicted results with real observations. In other words, it determines the best match between the estimated probability and the real probability (calculated using sample observations). Lower values of the AIC indicate a better compatibility of the model. The AIC was calculated for all four models: poison, negative binominal, poison zero inflated and negative binominal zero inflated. The "regular" Poison and negative Binomial models results depicted a better fit than obtained by the Zero Inflated models.

Detecting Significant Parameters

In order to identify the most significant parameters affecting road accidents, the GENMOD procedure using SAS software was applied on the "training set" (The GENMOD procedure fits generalized linear models, as defined by [24]. The class of generalized linear models is an extension of traditional linear models that allows the mean of a population to depend on a linear predictor through a nonlinear link function and allows the response probability distribution to be any member of an exponential family of distributions). Table 1 presents the results of executing the GENMOD procedure on the "500 m" segments data set. This procedure was also applied to identify such parameters on the models that were built for the 750, 1000, and 1500 m segment data sets.

Table 1. GENMOD procedure results for the training model (two-thirds of the data for segments of 500 m in length).

Parameter	Estimate	Standard Error	95% Confidence Limits		X^2	p
Intercept	−7.0910	0.6395	−8.3537	−5.8444	122.95	<0.0001
AADT	0.7144	0.0648	0.5883	0.8423	121.73	<0.0001
Area: South	−0.6788	0.1210	−0.9173	−0.4423	31.45	<0.0001
Area: Center	−0.5694	0.0890	−0.7444	−0.3952	40.94	<0.0001
Curvature	1.5948	0.4107	0.7953	2.4085	15.08	0.0001
Dispersion*	1.7419	0.1351	1.4925	2.0234		

As depicted at Table 1, the variables found to be correlated to accidents were: (1) *AADT*—annual average daily traffic volume on weekdays after a natural logarithm transformation; (2&3) Area—the events that occurred at the northern area, were analyzed and compared with the central and southern area (4) *Curvature*.

Examining the "dispersion" values of the training set shown in Table 1 (ranging from 1.4925 to 2.0234), it can be seen that the value "1" does not fall into the confidence interval at 95% significance level. Thus, it can be concluded that the data were characterized by over-dispersion. This indicates that the suitable model is a Negative Binomial.

Cross Validation—Evaluating the Models' Level of Compatibility

Expected Accidents—Real Accidents Validation

The first model's compatibility was determined by comparing the expected number of accidents (as calculated according to the model) with the real number of accidents at different road section lengths. Due to insufficient matching, the Empirical Bayes (EB) procedure was applied. EB method calculates the expected number of accidents, while taking into account the actual number of accidents that occurred on road sections [25]. The procedure was performed according to Equation (1) (presented by [11,26]).

$$\left(\frac{\lambda}{\frac{1}{\alpha}+\lambda}\right)\times\left(\frac{1}{\alpha}+y\right) \quad (1)$$

where:

α—over-dispersion parameter, as calculated for each one of the data sets (four different segment lengths)
λ—real number of accidents
y—predicted/expected number of accidents

Figure 1 presents the comparison results for the road section data set of 500 m in length, including the EB method. Figure 2 and Figure 3 present the breakdown of the original data into different observation numbers.

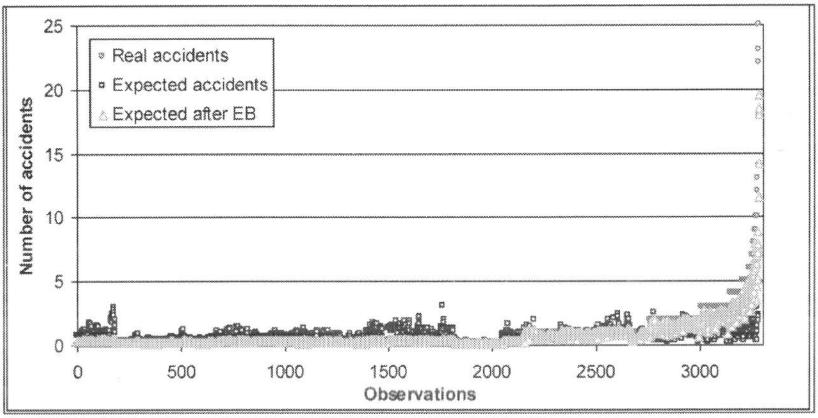

Figure 1. Expected number of accidents comparing real number of accidents and predicted number after applying EB method (road section of 500 m).

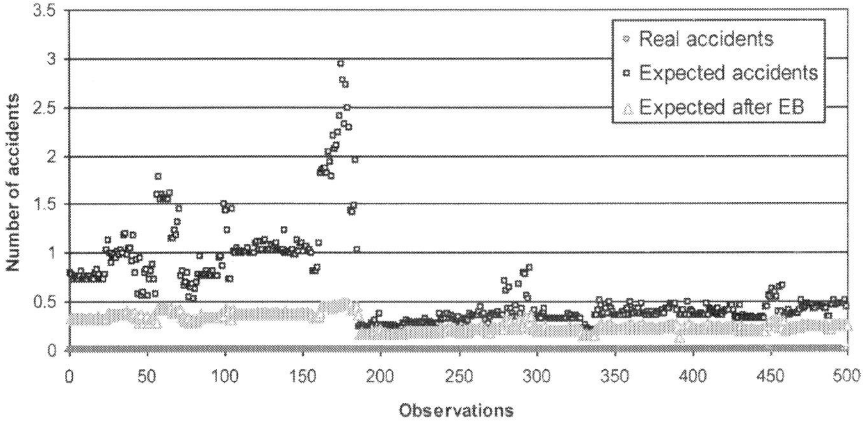

Figure 2. Expected number of accidents comparing real number of accidents and predicted number after applying EB method (road section of 500 m)—observation 0 until 500.

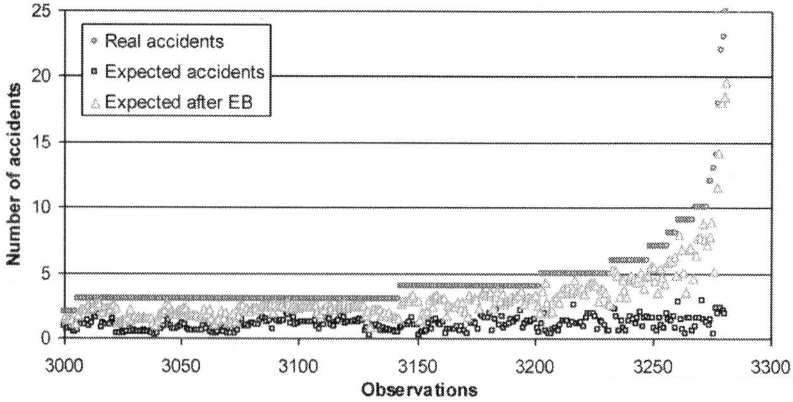

Figure 3. Expected number of accidents comparing real number of accidents and predicted number after applying EB method (road section of 500 m)—observation 3000 until 3300.

Comparison of Training and Control Models

The GENMOD procedure applied on the training model (two-thirds of the data) was also applied on the control model (one-third of the data). In advance, only the parameters that were found to be significant in the training model were integrated in the control model. Table 2 presents the coefficients of the control model.

Table 2. GENMOD procedure results for the control model (one-third of the data for segments of 500 m in length).

Parameter	Estimate	Standard Error	95% Confidence Limits		X^2	p
Intercept	−7.7780	0.9234	−9.6087	−5.9821	70.95	<0.0001
AADT	0.8019	0.0937	0.6197	0.9878	73.23	<0.0001
Area: South	−0.7366	0.1629	−1.0584	−0.4186	20.45	<0.0001
Area: Center	−0.8810	0.1249	−1.1274	−0.6369	49.75	<0.0001
Curvature	0.9894	0.5845	−0.1523	2.1489	2.86	0.0905
Dispersion	1.5877	0.1764	1.2701	1.9650		

Once the estimates were calculated for the two models, the coefficients were evaluated by testing whether the model coefficient estimates for the control model fall within the confidence interval of the training model at a confidence level of 95%. For example, in the control model (Table 2), the coefficient value of the variable representing traffic volume (*AADT*) is 0.8019. It can be seen that this value falls within the range of 0.5883 to 0.8423 (see Table 1), meaning that the model is statistically reliable enough for predicting this variable. From the comparison of Table 1 and Table 2, it can be seen that this is true for all variables in the model, except for the variable *Area: central* (referring to the difference between the central area and the northern area). Here, the coefficient value −0.8810 does not fall within the confidence interval of the training model (ranging from −0.3952 to −0.7444). This means that the model is not statistically reliable enough for predicting this parameter. In addition, one may see that the variable Curvature is not significant in the control model if incorporated into the model for its lowest AIC.

TOST Test for Equivalence

As stated above, it is possible to compare the expected values of accidents according to the training model (two-thirds of the data) and the expected number of accidents according to the control model (one-third of the data). In a utopian world, it would be expected that the mean difference between the two models would be zero, since both models predict the same phenomenon. However, in the real world, some expected differences between the means can be predicted to range around zero. Hence, one should ask only whether the differences are within a reasonable pre-defined range. For this purpose, was used the Two One-Sided Test (TOST), (T-testing for two independent populations is not enough to address this issue because T-tests assume the data to be normally distributed,as opposed to random events, such as car accidents). In addition, T-tests are used to prove unequal results, while here we need to prove precisely equally results) which allows to verify equality for any specified range. In this study, a minimal range of −0.25 to 0.25 accidents was defined for the evaluation. If the difference between the mean (μ) values falls within the confidence interval of the difference, it indicates that the model provides similar expected predictions. Equations 2 and 3 present the test hypotheses.

$$H_0 = \mu < \theta_L \; or \; \mu > \theta_U \rightarrow H_0 = \mu < -0.25 \; or \; \mu > 0.25 \tag{2}$$

$$H_1 = \theta_L \leq \mu \leq \theta_U \rightarrow H_1 = -0.25 \leq \mu \leq 0.25 \tag{3}$$

where:

θ_L, θ_U—upper and lower limit (±0.25 accident)
μ—mean difference

Table 3 presents the TOST test results for road sections of 500 m in length. If the values in column "90% CL Mean" are between the upper range (Upper Bound) and the lower range (Lower Bound), then we accept H0 and reject H1. This means that the values of expected accidents, as calculated for the two sets of data (one-third and two-thirds), are close enough.

Table 3. TOST results for 500 m segments.

Mean	Lower Bound		90% CL Mean			Upper Bound	Assessment
−0.0548	−0.25	<	−0.0832	−0.0264	<	0.25	Equivalent

Proportions of Probabilities Comparison

It is optional to calculate estimations for the probability of the occurrence of zero accidents on road segments, one accident, two accidents, and so on. If the model is reliable enough, then we should obtain similar proportions while comparing the percentage of sections in which there were "zero accidents" in practice (using "real" world, "real" data) and the average probability of expected "zero accidents", as calculated by the model.

Table 4 presents the calculated probabilities for the occurrence of N accidents on road sections of 500 m, as calculated from the model. This is presented for a sample of 9 segments. Each row in the table represents a specific road section of a specified length (as stated, 500 m in the current example). Each column represents the probability of the occurrence of N accidents on that section. Thus, the PO0 column represents the probability of the occurrence of zero accidents; the PO1 column represents the probability of the occurrence of one accident, and so on.

Table 4. Probabilities for the occurrence of N accidents for a sample of sections of 500 m in length.

PO0	PO1	PO2	PO3	PO4	PO5	PO6	PO7	PO8
0.610277337	0.2021285764	0.0917807833	0.0454338563	0.023412922	0.0123615103	0.0066255263	0.0035899153	0.0019608
0.60239497	0.2027937693	0.0935947405	0.0470925361	0.0246749411	0.013237001	0.0072112615	0.0039714387	0.0022048
0.606145977	0.2024918782	0.0927388184	0.0463040849	0.0240757855	0.0128165475	0.0069286689	0.0037865486	0.0020861
0.62202776	0.2009173652	0.0889710565	0.0429520422	0.0215934632	0.0111145072	0058096008	0.0030698513	0.001635
0.614599875	0.2017137307	0.090761575	0.0445217661	0.0227429164	0.0118945867	0.0063174341	0.0033919314	0.001835
0.608603762	0.2022796466	0.0921707895	0.0457865823	0.026857182	0.0125448166	0.0067473035	0.0036686911	0.0020109
0.596272816	0.2032296575	0.0949626287	0.0483751326	0.0256688671	0.0139378986	0.0076875464	0.0042864047	0.0024093
0.620124035	2.2011315691	0.0894345832	0.0433545348	0.0218860298	0.0113117258	0.0059371623	0.0031502422	0.0016850
0.622048000	0.2009150508	0.0889661124	0.0429477634	0.021590361	0.0111124209	0.0058082544	0.0030690047	0.001634

Table 5 presents the mean of columns PO0 and PO1 (the average probability for zero accidents, the average probability of one accident, and so on), as presented in Table 4.

Table 5. Mean probabilities for the occurrence of zero and one accidents on 500 m segments.

Variable	N	Mean	Std Dev	Minimum	Maximum
PO0	2188	0.6582303	0.1082179	0.3312983	0.8639541
PO1	2188	0.1837504	0.0245488	0.1115330	0.2043939

Table 6. Real data accident proportions for N accidents (road segments of 500 m in length).

AccCNT				
AccCNT	**Frequency**	**Percent**	**Cumulative Frequency**	**Cumulative Percent**
0	1441	**65.86**	1441	65.86
1	397	**18.14**	**1838**	**84**
2	170	7.77	2008	91.77
3	87	3.98	2095	95.75
4	37	1.69	2132	97.44
5	25	1.14	2157	98.58
6	12	0.55	2169	99.13
7	5	0.23	2174	99.36
8	2	0.09	2176	99.45
9	4	0.18	2180	99.63
10	3	0.14	2183	99.77
13	1	0.05	2184	99.82
14	1	0.05	2185	99.86
18	1	0.05	2186	99.91
23	1	0.05	2187	99.95
25	1	0.05	2188	100

The "PO1" column in Table 5, presents the model calculated probability for one accident occurrence for all database segments. The mean value was 0.1837 (18.37%). This value is very close to the value of 18.14%, which is the value of the actual proportion of segments that had only one accident on them, in the 500 m length database during the study period (see Table 6 at column AccCNT = 1).

Training Model (Two-Thirds Data) and Full Control Model (One-Third Data) Comparison

A new model based on one-third of the data was constructed to serve as the "full control model" due to its use of all existing parameters in the data set, as opposed to the previous "control" models in which only the significant parameters were combined. The comparison between the two models provided an additional control and enabled an assurance test on the results of the training model (two-thirds of the data) at various levels. It also allowed for validation of the model type (data distribution aspect). Thus, it served as an additional control for the parameters obtained from the significant model, as well as a control for the estimates obtained from the model.

RESULTS, ANALYSIS AND DISCUSSION

Optimal Models

The study shows that the shorter segment data sets tested (lengths of 500 and 750 m) yielded the most stable models. This means that recommendations can be made concerning investments in the infrastructure and enforcement for short sections, as opposed to the prevailing attitude of improving relatively long sections of the road network.

Equation (4) represents the expected rate of road accidents (μ) for highway sections of 500 m in length:

$$\mu = e^{-7.091 + 0.714 \times AADT\_Log + (-0.678) \times \text{Region1} + (0.569) \times \text{Region3} + (1.594) \times CurvePrm3} \quad (4)$$

Equation (5) represents the expected rate of road accidents (μ) for highway sections of 750 m in length:

$$\mu = e^{-5.9688 + 0.6420 \times AADT\_Log + (-0.7316) \times \text{Region1} + (0.5483) \times \text{Region3} + (1.0741) \times CurvePrm3} \quad (5)$$

where: *AADT_Log*—annual average daily traffic after it has been transformed using a natural logarithm. Region—while Region1 = the southern area; Region 2 = the northern area; and Region 3 = the central area, the model compares and analyzes the northern region in relation to the two other regions. Thus, for a section located in Region 2, the value that will be placed at variables Region 3 and Region 1 will be 0. Curvature—segment curvature. Calculated by the sum of the differences between the nearest H, divided by segment length (H in red: the vertical distance between the parabola tip and the line segment connecting the strands—appears as a dotted line), as specified in Equation (6) and Figure 4.

$$Curve = \frac{\sum |H_{i+1} - H_i|}{Total\_Length} \tag{6}$$

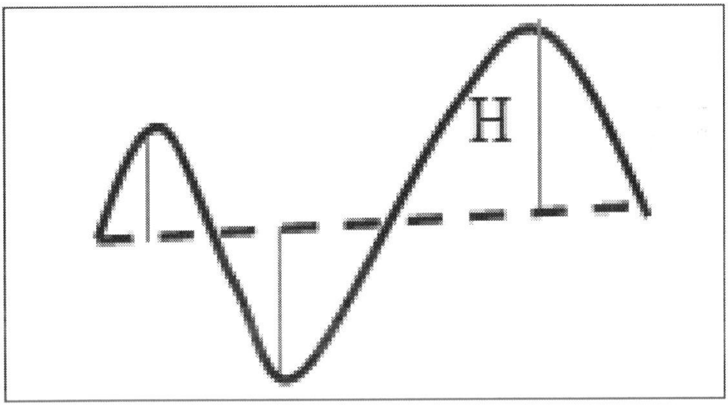

Figure 4. Curvature calculation.

After calculating the expected accident rates, the EB method was applied in order to further improve the model.

Model Parameters

Of all the parameters that were examined in this study, the curvature of the road section, the region, and the traffic volume were found to be the most significant in terms of affecting accidents. Due to its high expected accident rates the northern region was designated for special observation and was divided into sub-regions that were analyzed separately.

Curvature

The curvature and traffic volume findings match [18] study. However, unlike Miaou's research, which identified a relationship between curvature and the accident rate only for large trucks, the current study addressed all types of vehicles and carried out rules for the data using the EB method.

The traffic volume (*aadt log*) parameter was also found to be meaningful. This finding matches other studies, including [5,6,10,27].

Region

The third variable found to be significant was the region. In Israel, this variable is compatible with the major divisions of the "periphery" (North and South) and the central area (Tel Aviv metropolis). All of the analyses consistently showed the highest expected accident rates to be in the northern area and particularly in the Western Upper Galilee. Therefore, the results concerning this variable should be further analyzed and checked in several aspects.

The region variable was analyzed during the first step of building the models. This parameter appeared to be significant in all four models, as constructed for various lengths of road sections. In all four models, the accident rate for the northern part of Israel was expected to be higher (approximately twice as high) than the expected rate of accidents for the central and southern parts of Israel.

The second regional analysis examined the statistical differences between the northern, southern, and central regions. This analysis aimed to examine the statistical differences between the regions, and to learn whether a particular area will depict more or less accidents, regardless of the volume of traffic. The data set was divided into several traffic volume ranges: "low", "medium-low", "medium-high", and "high". Due to the need to maintain balanced groups for the statistical analysis, only groups with sufficient samples were analyzed. In practice, only the "medium–low" group (5000 to 15,000 vehicles per road section) and the "medium–high" group (15,000 to 30,000 vehicles per road section) were analyzed.

In this analysis, the trends were similar at road lengths of 500 and 750 m for the "medium–low" and the "medium–high" road sections, with the expected accident rate in the northern region greater than expected in the southern region and the expected accident rate in the northern region greater than expected in the central region. These trends are also consistent with the results obtained in the previous regions analysis performed on the training model (the expected accident rate at north is higher than expected at the central and southern regions.

When comparing the predicted accident rates of the central and southern regions, different relations were found:

1) At "medium-low" traffic volumes, the expected accident rates in the central region were greater than expected in the southern region. This may be explained by the fact that in the south, despite dealing with accidents at so-called medium-low traffic volumes, the traffic volumes are truly lower, meaning that the roads actually carry very sparse traffic. The logic is simple: Given that very few vehicles (low range value) are present, the probability for accidents drastically decreases in this region, as compared to the central region, where vehicles are present at the higher end of the "medium-low" traffic volume range.
2) At "medium-high" traffic volumes, the expected accident rates in the southern region were greater than expected in the central region. The presumed explanation is that as traffic increases, it is more vulnerable to the effects of the infrastructure quality, such as unregulated intersections that flow directly into the highways, fewer lanes, and lack of lane separation. As the infrastructure in the southern region was much poorer than in the central region at the time of the study, it affected the accident sensitivity rates.

The third aspect of the regional analysis was the interaction analysis. Unlike the previous analyses, here, the traffic volume serves as a parameter. Thus, there is difference in division of the traffic volume thresholds. This was carried out by partitioning the segments into three groups: "low" traffic volume, characterized by values ranging from 0 to 10,766; "medium" traffic volume, characterized by values ranging between 10,767 and 20,033; and "high" traffic volume, characterized by values over 20,033.

An interaction between the region and the traffic volume variables was found for both data sets examined. This means that the effect of traffic volume on the expected rate of accidents is not uniform and depends on the region as well (see Table 7). The interaction analysis findings match the results of the first and second analyses. All other interactions such as region and curvature, and region and slope were not found significant, and thus, were not further examined as were the AADT and region. The significance test was carried out by comparing AIC value of "Null model" (The "Null Model" containing the variables: curvature, area and volume of traffic without variable representing interaction), with three other models that contained also interactions as parameters.

The anomaly as depicted at Table 7 where the expected results for "Medium traffic" is higher than for the "High" in the southern area is intriguing indeed. This is biased due to the division of the traffic groups. The maximum daily traffic in the south area was 23,000. This further implies that the "High traffic" group almost does non exit in this area. Only 72 records were found in the south area for this group and this actually biased the results.

Table 7. Interactions—number of expected accidents divided into regions (Section lengths of 500 m and 750 m).

Data Set	AADT	Area	Expected Number of Accidents
	Low	south	0.51
	medium	south	1.51
	High	south	1.19
	Low	north	1.64
500 m	medium	north	2.29
	High	north	4.10
	Low	center	1.32
	medium	center	1.05
	High	center	2.38
	Low	south	0.74
	medium	south	2.3
	High	south	1.89
	Low	north	2.22
750 m	medium	north	3.55
	High	north	5.45
	Low	center	2.06
	medium	center	1.41
	High	center	3.74

In the last analysis of the region variable, the northern area was designated for special observation due to its high expected accident rates. The area was divided into eight sub-regions, and the analysis was performed on the full accident data (without dividing into 2/3 and 1/3). One sub area, "Golan Heights", was eliminated from the analysis due to low number of accidents events during the research period. The Linear Step-up procedure was applied in order for detecting false discovery rate. The analysis results showed significant high expected accident rates in the Western Upper

Galilee sub-region. According to this finding, it is recommended that a range of practical activities be carried out, from improvement of infrastructure to education and advocacy.

Speed

Interestingly, the speed variable was not determined to be significant by this study. This conclusion is in agreement with other published studies, such as [15], which found that the accident rate rose in accordance with increasing speed on non-major roads. Yet, on main roads, it was found that the path width, the density of nodes, and traffic had higher correlations with the accident rate.

The speed parameter, as integrated into this study, was only estimated. In order to be able to assess this parameter more accurately in future studies, it is necessary to acquire and integrate more accurate speed data, perhaps even from different points in time, such as day and night, weekdays and weekends, and so on.

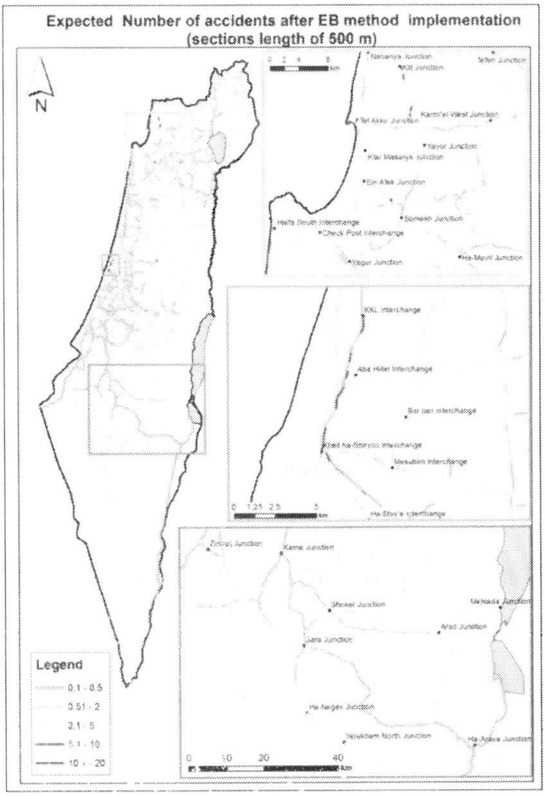

Figure 5. Expected number of accidents (road sections at length of 500 m).

Solar Angle (Sun Glare)

Each accident was positioned using ESRI ArcGIS software address matching process. After the location was found the azimuth and slope of the car were calculated based on Digital Elevation Model.

The sun glare value was calculated for each accident using a batch of equations migrated to MS Access software. Due to the high uncertainty of the direction of the car during the accident (upstream or downstream) many of the resulted declination were incorrect. This affected the integrity of the sun glare values. Thus the sun glare was taken out as a parameter in this research but it is recommended to use it once we have accurate and valid data.

Prediction Maps

Figure 5 and Figure 6 present some of the final research outputs. The maps display the expected number of accidents on road sections after the correction procedure of EB. The amplitude of the expected number of accidents is depicted by a color palette, ranging from green, which represents road sections with a low number of expected accidents, to red and purple, representing sections with high rates of expected accidents. It should be emphasized that when positioning the accidents on the road sections, the direction was selected according to CBS data despite its lack of precision [23]. The rationale here was that in terms of physical road parameters, it can be assumed that dangerous road sections from one direction would also be dangerous on the opposite lane.

Figure 6. Expected number of accidents (road sections at length of 750 m).

REPRODUCIBILITY AND OPPORTUNITIES FOR FURTHER STUDY

This paper described the development of GIS-based prediction models for the assessment of traffic accidents in highway segments, using spatial and traffic-based parameters. The findings of this research constitute a basis and a starting point for other varied studies in the field of road safety. First and foremost it is proposed to examine the quality of the model on accident data of future years. This model can also be implanted in other countries/regions and in addition, at various segmentation e.g., roadway functional classes, vehicle configurations, types of crashes (e.g., those involving drunk drivers), and crash severity levels (e.g., fatal, injury, and non-injury events). It is also suggested to perform spatial autocorrelation, looking into influence of nearby segments.

According to the research finding, the analyses of the "region" variable produced very distinct and alarming high expected accident rates in the northern region. This is distinct especially in the western Upper Galilee sub-region. Also remarkably distinct were the substantiated findings of interaction between the "region" and the "traffic volume" variables. The conclusion is that the effect of "traffic volume" on the expected rate of accidents is not uniform and depends on the region as well. Thus, suggested is to investigate the "region" and the "sub region" variables more closely while associating: environmental, socioeconomic and economic variables to the model. Among these are: infrastructure investments, shoulders width, lanes separation, number of lanes, *etc.*

The speed parameter, as integrated into this study, was only estimated. In order to assess this parameter more accurately in future studies, it is necessary to acquire and integrate more accurate speed data. Hence, suggested here is getting speed data from navigation applications and traffic reports based on user's community and to analyze speed effect at different times (day and night, weekdays and weekends, and so on).

We believe that the Scholl width and number of lanes are significant variables of the existence of road accidents. These data is not collected currently at continuously and accurately as it should. Accurate data for both these variables will outcome with more reliable results and models. In addition, accident orientation data which is considered as not reliable at this study is also recommended to integrate while placing accident at its occurrence site in other studies.

CONCLUSIONS

This paper described the development of a GIS-based prediction model for the assessment of traffic accidents in highway segments, using spatial and traffic-based parameters.

The most significant and practical "revelation" in this study is that the short road sections yielded the most stable models. This allows aiming the treatments on short sections of the road as a way to save resources and potentially gain maximum benefit at minimum investment. The ability to narrow down the high-risk areas to specific road sections also enable to focus the police (and other) enforcement teams at those sections in order for save life. The models were improved by using the Empirical Bayes method, which increased the accuracy of the assessment by taking into account historical data and correcting the biases from the "regression to the mean" phenomenon. In addition, it was found that the most significant variables affecting accident rates were: curvature of the road section, the region, and the traffic volume. In addition, an interaction between the region and the traffic volume variables was found. This means that the influence of traffic volume on the expected rates of accidents is not uniform and depends on the region as well. Investigating this issue more profoundly will probably raise some very interesting finding. The categorization of the data set according to road sections of different but equal lengths allowed "neutralizing" the "segment length" parameter, which in many studies was found to be a significant parameter in the model.

This study confirms, once again, the power and benefits of using Geographic Information Systems (GIS) in the field of road safety (GIS-T). This approach allows for displaying and examining the data in "real" space (*i.e.*, the physical location where the events occur), iteration, and performing complex statistical-spatial analysis in a relatively simple manner.

ACKNOWLEDGMENTS

This research was funded by the Ran Naor Foundation established by Or Yarok Association for the advancement of road safety research, Grant No. MK-020-2008. The authors wish to thank Pavel Goldshtein, from the statistics consulting unit at the University of Haifa for his support in the statistical analyses and to Basheer Haj-Yehia for his contribution to the GIS processing.

AUTHOR CONTRIBUTIONS

This manuscript is written based on the dissertation of Keren Pollak at the University of Haifa, Ammatzia Peled served as the PhD supervisor. Shalom Hakkert was an expert consultant for transportation. All authors read and approved the final manuscript.

REFERENCES

1. Page, Y. A statistical model to compare road mortality in OECD countries. *Accid. Anal. Prev.***2001**, *33*, 371–385.
2. Miaou, S.P.; Song, J.J.; Mallick, B.K. Roadway traffic crash mapping: A space-time modeling approach. *J. Transp. Stat.* **2003**, *6*, 33–57.
3. Peled, A. *Detection Urban Safety Issues Using Geographic Information System (GIS)*; Institute of Transportation Studies: Jerusalem, Israel, 1996.
4. Hakim, S.; Shefer, D.; Hakkert, S.; Hocherman, I. A critical review of macro models for road accidents. *Accid. Anal. Prev.* **1991**, *23*, 379–400.
5. Levine, N.; Kim, K.; Nitz, L. Spatial analysis of Honolulu motor vehicle crashes: I. Spatial patterns. *Accid. Anal. Prev.* **1995**, *27*, 663–674.
6. Levine, N.; Kim, K.; Nitz, L. Spatial analysis of Honolulu motor vehicle crashes: II. Zonal generators. *Accid. Anal. Prev.* **1995**, *27*, 675–685.
7. Summala, H. Accident risk and driver behavior. *Saf. Sci.* **1996**, *22*, 103–117.
8. Jones, A.P.; Langford, I.H.; Bentham, G. The application of K-function analysis to the geographical distribution of road traffic accident outcomes in Norfolk, England. *Soc. Sci. Med.***1996**, *42*, 879–885.
9. Beenstock, M.; Goldin, A.; Gafni-Sri, D. *The Relationship between Traffic Enforcement and Road Accidents: Statistical Analysis*; Office of the Chief Scientist, Ministry of Public Security: Jerusalem, Israel, 1998.
10. Oppe, S. Development of traffic and traffic safety: Global trends and incidental fluctuations.*Accid. Anal. Prev.* **1991**, *23*, 413–422.
11. Ng, K.S.; Hung, W.T.; Wong, W.G. An algorithm for assessing the risk of traffic accident. *J. Saf. Res.* **2002**, *33*, 387–410.
12. Schneider, R.J.; Ryznar, R.M.; Khattak, J.A. An accident waiting to happen: A spatial approach to proactive pedestrian planning. *Accid. Anal. Prev.* **2004**, *36*, 193–211.

13. Qin, X.; Ivan, J.N.; Ravishanker, N.; Liu, J. Hierarchical Bayesian estimation of safety performance functions for two-lane highways using Markov chain Monte Carlo modeling. *J. Transp. Eng.* **2005**, *131*, 345–351.
14. Sabel, C.E.; Kingham, S.; Nicholson, A.; Bartie, P. *Road Traffic Accident Simulation Modelling—A Kernel Estimation Approach*; Otago University: Dunedin, New Zealand, 2005.
15. Aarts, L.; van Schagen, I. Driving speed and the risk of road crashes: A review. *Accid. Anal. Prev.***2006**, *38*, 215–224.
16. Song, J.J.; Ghosh, M.; Miaou, S.; Mallick, B. Bayesian multivariate spatial models for roadway traffic crash mapping. *J. Multivar. Anal.* **2006**, *97*, 246–273.
17. Miaou, S.P.; Lum, H. Modeling vehicle accidents and highway geometric design relationships.*Accid. Anal. Prev.* **1993**, *25*, 689–709.
18. Miaou, S.P. The relationship between truck accidents and geometric design of road sections: Poisson *versus* negative binomial regressions. *Accid. Anal. Prev.* **1994**, *26*, 471–482.
19. Baruya, A. Working Paper: A Review of Speed-Accident Relationship on European Roads. Master's Thesis, Transport Research Laboratory, Wokingham, UK, August 1997.
20. Aguero, V.J.; Jovanis, P.P. Analysis of road crash frequency with spatial models. *Transp. Res. Rec.* **2008**, *2061*, 55–63.
21. Peled, A.; Hakkert, A.S. A PC-oriented GIS application for road safety analysis and management.*Traffic Eng. Control* **1993**, *34*, 355–361.
22. Peled, A.; Haj-Yehia, B.; Hakkert, A.S. ArcInfo-Based Geographical Information System for Road Safety Analysis and Improvement. Available online: http://proceedings.esri.com/library/userconf/proc96/to50/pap005/p5.htm (accessed on 13 August 2013).
23. The Central Bureau of Statistics. *Road Accidents with Injuries in 2007 Part II: Accidents on—Rural Roads*; The Central Bureau of Statistics: Jerusalem, Israel, 2007.
24. Nelder, J.A.; Wedderburn, R.W.M. Generalized linear models. *J. Royal Stat. Soc. Ser. A* **1972**,*135*, 370–384.
25. Hauer, E. Empirical bayes approach to the estimation of "unsafety": The multivariate regression method. *Accid. Anal. Prev.* **1992**, *24*, 457–477.
26. Sayed, T.; Rodriguez, F. Accident prediction models for urban unsignalized intersections in British Columbia. *Transp. Res. Rec.* **1999**, *1665*, 93–99. [Google Scholar]
27. Hauer, E. Overdispersion in modeling accidents on road sections and in Empirical Bayes estimation. *Accid. Anal. Prev.* **2001**, *33*, 799–808.

CITATION

Keren Pollak, Ammatzia Peled and Shalom Hakkert, Geo-Based Statistical Models for Vulnerability Prediction of Highway Network Segments, doi:10.3390/ijgi3020619.

CHAPTER 2

Stormwater Field Evaluation and Its Challenges of a Sediment Basin with Skimmer and Baffles at a Highway Construction Site

Xing Fang[1], Wesley C. Zech[1] and Christopher P. Logan[2]

[1]Department of Civil Engineering, Auburn University, Auburn, AL 36849-5337, USA;

[2]Turner Construction Company, 5300 Virginia Way, Brentwood, TN 37027, USA

ABSTRACT

A field-scale data collection plan to monitor and evaluate the performance of a sediment basin design was developed and implemented using portable automatic stormwater samplers, flow modules, a rain gauge, and inflow weirs. The design configuration consisted of a skimmer as the primary dewatering device, three coir baffles installed inside the basin, polyacrylamide flocculant blocks and ditch checks in the inflow channel. A sediment basin built on a highway construction site in Franklin County, Alabama, U.S. using the aforementioned design configuration was monitored over 16 rainfall events from 15 November 2011 to 6 February 2012. The basin effectively removed sediments during the early stages of construction when the correct type of polyacrylamide flocculant blocks was used, e.g., 97.9% of sediment removal after a rainfall event on 16 November 2011. It is difficult and challenging to dose sediment-laden stormwater inflow with an exact amount of flocculating agent across all runoff producing events since rainfall is a stochastic variable. Based upon results from this study, it is recommended that a minimum volume of 251.9 m^3/ha of contributing drainage area be used to sufficiently size a basin, which is still significantly under-designed for a 2-year, 24-h storm event in the southeast.

This paper presents challenges and lessons learned regarding sediment basin design, monitoring, and performance that are beneficial to future studies.

INTRODUCTION

Sediment-laden stormwater runoff from poorly managed construction sites is one of the leading causes of nonpoint source (NPS) pollution in the U.S. and over the world [1]. Construction activities (*i.e.*, land clearing, excavating, and grading) expose bare soil to forces of wind, rain, and stormwater runoff, greatly increasing the potential for erosion. According to the United States Environmental Protection Agency (USEPA), sediment entrained in construction-site stormwater runoff is one of the most widespread pollutants affecting rivers and streams, second only to pathogens [1]. Construction-site sediment yields can be 10 to 20 times greater than those from agricultural lands, and about 1000 to 2000 times greater than those from forested lands [1]. These concerns coupled with growing pressure from regulatory agencies and public scrutiny have resulted in demands to provide enhanced methods for controlling erosion and preventing sediment-laden stormwater discharges from construction sites. Although the USEPA withdrew the federal numeric effluent limitation [2], many state environmental agencies enforce numeric effluent limitations. For example, the Alabama Department of Environmental Management (ADEM) requires effluent turbidity to be less than 50 nephelometric turbidity units (NTU) above background levels for any Alabama waterbody with fish and wildlife designated uses [3]. If stormwater runoff from a construction site violates the effluent limitation, the discharge is prohibited and the construction site operator has to remediate the problem [3].

Various NPS pollution abatement methods and best management practices (BMPs) are commonly used to reduce erosion and sediment transport prior to sediment-laden runoff discharging into a receiving water body. One method of preventing sediment from migrating from construction sites is by capturing it in a sediment basin [4]. Sediment basins provide stormwater detention, promoting gravitational and chemically assisted settling. In this case study, sediment basin monitoring protocols were developed to monitor and evaluate a sediment basin design configuration adopted by the Alabaman Department of Transportation (ALDOT). The sediment basin monitored was constructed on an ALDOT highway construction site in Franklin County, Alabama (AL), U.S. In the past, ALDOT has used a traditional sediment basin design adapted from the 2006 Alabama Handbook [5]. Many traditional sediment basin designs use a 45.7 cm diameter perforated riser pipe as the primary outlet structure [6]. Routine stormwater inspections following rainfall events have revealed that perforated riser pipes do not allow sediment-laden stormwater to be detained for a sufficient period of time. Furthermore, the riser structure allows basin dewatering to occur across the entire depth of the water column. Following 2009 Alabama Handbook design guidance [7], the ALDOT basin evaluated in this study used a floating surface skimmer as the primary dewatering device with a rip-rap armored emergency spillway to control discharge during extreme rainfall events (Figure 1). The basin design also used three coir baffles positioned inside the basin, polyacrylamide (PAM) application at the inflow channel, as well as an excavated sump, a rip-rap ditch check, and a rip-rap lined inflow channel (Figure 1). The basin design with a skimmer and baffles was based on many years of research performed by the Pennsylvania State University [8,9,10] and the North Carolina State University [11,12,13,14], along with experiences of the North Carolina Department of Transportation [15] and other stormwater management agencies in employing these practices on active construction sites.

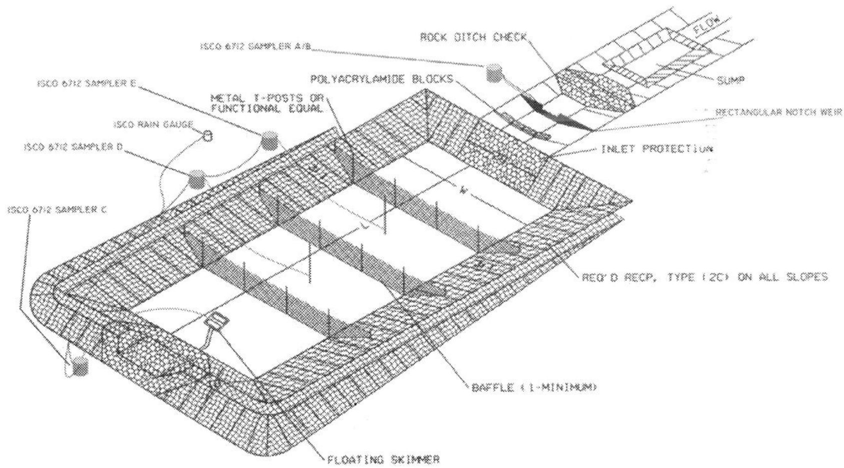

Figure 1. Isometric view of the sediment basin layout including data collection equipment setup.

There are several published field studies on monitoring the performance of sediment basins [13,16,17] and some small- to medium-scale performance-based experimental research studies on sediment basins [6,8,9,10,18,19]. Millen *et al.* [8] reported that a floating surface skimmer discharged the least total sediment in comparison to four sediment/water control treatments that included a perforated riser in an experimental sediment basin. Bhardwaj and McLaughlin [19] found that both active and passive PAM dosing systems significantly reduced turbidity. McLaughlin *et al.* [17] reported that basin modifications (e.g., using baffles and a skimmer) have less of an effect on turbidity improvement in comparison to the addition of PAM. The effectiveness of a sediment basin to remove suspended sediments depends on various factors such as: Sediment particle size, ratio of basin surface area to inflow rate [20], sediment properties (e.g., particle density, shape, and concentration), and particle settling velocities. Sediment basins have shown high removal efficiencies for suspended solids, and for heavy metals and organic compounds that have attached to soil particles [21]. Some studies show that a sediment basin can trap more than 75% of incoming sediments [22]. A more recent study however, showed sediment control systems removed only 35% to 60% of incoming sediments from highway construction sites [17]. The removal efficiency of sediment control devices depends on sediment-laden runoff characteristics, which are affected by intensity and duration of storm events, topography and extent of construction limits, soil type and characteristics, vegetative cover composition, and other BMPs implemented upstream [23].

Sediment basins remove suspended solids from stormwater by gravitational settling [24]. A single particle in clear, quiescent water can eventually settle with a constant velocity [25], this is known as the terminal velocity. According to Stokes' Law [26], settling velocity is directly related to particle characteristics such as shape factors, specific gravity, and viscosity (function of temperature) [25]. Stormwater runoffs on construction projects often contain very fine suspended sediments that are too small to gravitationally settle under normal conditions. Smaller particles are more susceptible to resuspension due to their position on the basin floor (e.g., last to settle), as well as size and mass [27]. Bhardwaj *et al.* [28] observed that neither total suspended solids (TSS) nor turbidity was reduced in an open sediment basin (with no baffles) at 1.5-h and 24-h detention times, suggesting that suspended materials were very resistant to settling. It has been determined that clay and silt fractions are the greatest contributors to increased turbidity in runoff [19]. Chemical treatments using coagulants or flocculants, such as PAM, promote the process of suspended sediment bonding to each other to enhance settling. PAM is a high molecular weight synthetic polymer that can be manufactured to have a variety of chain lengths, and to have an anionic, nonionic, or cationic net charge [19]. The preferred form of PAM is the anionic form due to its low aquatic toxicity. Anionic PAM binds to suspended sediment largely through rapid and irreversible cation bridging, pulling particles together into flocs, resulting in an effective chemical flocculant [19].

Zech *et al.* [29] conducted a survey to determine the state-of-the-practice for sediment basin design, construction, maintenance, and inspection techniques employed by state highway agencies nationwide in the U.S. The survey showed that few agencies actually monitor or collect data from sediment basins. Therefore, performance characteristics for various design configurations of sediment basins remain unknown. There is a lack of information and guidance detailing monitoring protocols for a sediment basin to collect data and analyze the performance of basins. The objective of this paper is to summarize the monitoring data collection plan developed and implemented, along with results of the data analyses performed in a basin constructed on a highway construction site. Even though performance data in a single basin in a unique region may have limitations to be applied to other basins or regions, lessons learned and challenges in monitoring an active sediment basin are also presented. Sharing the experiences (*i.e.*, lessons learned and challenges) and results developed during this study will provide beneficial information to the professional community that may engage in monitoring sediment basin performance. Recommendations were also developed and can be used to enhance the performance of sediment basin designs and implementation. In general, the study is related to monitoring and engineering design of sediment basins used as a sediment control feature for construction stormwater management. The purpose of a sediment basin is to capture suspended solids in an effort to improve the water quality of stormwater runoff and to protect downstream water resources.

DATA COLLECTION PLAN

A field-scale data collection plan to monitor and evaluate the performance of a sediment basin was developed and implemented for the case study and consisted of the following elements: (1) rainfall and flow monitoring; (2) stormwater runoff sampling; (3) turbidity and TSS analysis of stormwater runoff samples; (4) soil or stormwater samples to determine particle-size distributions; and (5) a pre and post sediment basin survey to determine the volume of sediment deposited. The data collection plan was adapted and revised from the strategy used for experimental studies of a medium-scale sediment basin at the Sediment and Erosion Control Research and Education Facility in Raleigh, North Carolina [19,28].Figure 1 illustrates the sediment basin with a single inflow channel along with all necessary structural components and sampling equipment used for data collection. The basin design uses a floating surface skimmer as the primary dewatering device, three baffles installed within the basin, and PAM flocculant (floc) blocks placed within the inflow channel (Figure 1). ALDOT design specifications state that baffles shall be constructed of 100% coconut (coir) fiber materials supported between posts with a wire mesh backing and show that the installed baffle height shall protrude above the spillway crest elevation. The traditional sediment basin design used by ALDOT and many other agencies used a 45.7 cm diameter, perforated riser pipe as the primary outlet structure [5,6]. To monitor water quality in reference to sediment basin performance, five ISCO 6712 (Teledyne Technologies, Inc., Thousand Oaks, CA, USA) full-size portable stormwater samplers, shown in Figure 1, were used to collect stormwater samples at the following locations: Inflow (Sampler A or B), within the sediment basin (Samplers D and E), and outflow (Sampler C).

Rainfall and Flow Monitoring

In an effort to collect rainfall and inflow data, several data collection units were deployed. A tipping bucket rain gauge was connected to Sampler C, shown in Figure 2a,b, to monitor rain events on-site, giving accurate time stamped information regarding rainfall amounts and intensity. Figure 2b shows the basin after a rainfall event when both primary and secondary inflow channels were in place. Upstream of the sediment basin, rectangular notched weirs were installed in each channel to gauge inflow into the basin as shown in Figure 1. To properly accommodate inflow from both channels, the rectangular notched weirs were cut to 0.91 m and 1.22 m width for secondary and primary inflow channels, respectively, and 0.46 m in depth. The purpose of these weirs was to provide a means for determining the volume of stormwater inflow entering the basin as well as to conveniently collect inflow water quality samples using the ISCO 6712 samplers.

A 730 bubbler flow module (Teledyne Technologies, Inc., Thousand Oaks, CA, USA) was used to continuously collect necessary data to determine inflow discharges and volumes whenever runoff from the contributing drainage area flowed over the weir. The bubbler flow module was pre-programed and pre-calibrated with the dimension of the rectangular weir, based on the mounting depth below the base of the weir opening, so that as water flowed through the weir it registered a positive water level.

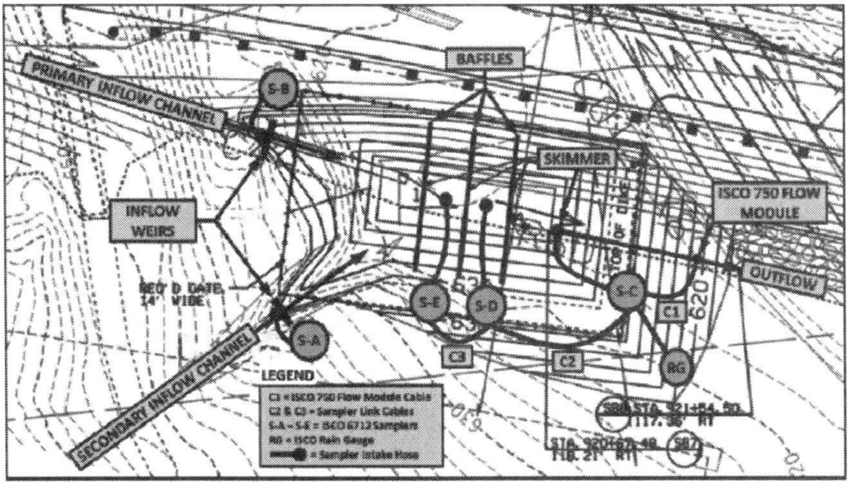

(a)

Figure 2. (**a**) Plan view (schematic diagram) and (**b**) a photo of actual project sediment basin layout and equipment locations, showing the condition when the water level was below the baffle height, and (**c**) the basin condition when water level breached above installed baffle height.

Stormwater Runoff Sampling

To understand the performance of the sediment basin, it was important to collect stormwater runoff samples at various locations to analyze for turbidity and TSS levels. Inflow sampler A was programmed to take 0.25 L stormwater runoff samples for every 1.42 m^3 of inflow passing over the weir in the secondary inflow channel. Inflow Sampler B was programmed to take a 0.25 L sample for every 4.25 $m^3$of inflow passing over the weir in the primary inflow channel, since inflow volumes were much larger in comparison to the secondary inflow channel.

Due to variations of depth, intensity, and duration of rainfall events and variations in the contributing drainage area of the construction site, runoff rates, volumes, and durations varied greatly from one rainfall event to another. Due to these variances, it was difficult to program the samplers to collect adequate inflow samples for different rainfall events to characterize inflows into the basin. ISCO 6712 samplers in the primary and secondary inflow channels were set to collect samples at fixed inflow volumes passing over the weirs rather than at fixed time intervals. This protocol was decided because inflow rates were unsteady and dependent on individual rainfall characteristics. The determination to collect stormwater samples for every 1.42 and 4.25 m^3 of inflow passing over the weirs was determined based on historical rainfall events that occurred in the study area and runoff estimations; a function of the contributing area and rainfall event.

For each inflow sampler, four 0.25 L samples were collected in a single 1 L bottle, creating a composite stormwater runoff sample to provide a measure of incoming water quality over the course of a rainfall event. In total, each of the inflow samplers had the capability of collecting up to 96, 0.25 L samples in a single program spanning a single rainfall event with a combined maximum inflow volume of 543.7 m^3. The ideal method to program a sampler to capture adequate and representative samples is to adjust the sampler program based on runoff modeling and rainfall forecasts so that an adequate number of inflow samples are always collected for various rainfall events with small and large rainfall depths. However, this is difficult to implement due to the lack of rainfall forecast information and forecast accuracy for a particular area. Typically, weather forecasts provide the probability of having a rainfall event without projecting rainfall depth and duration. In this study, the programs for the ISCO 6712 samplers were fixed for all 16 events monitored. This type of programming resulted in collecting minimal inflow samples for events with relatively small rainfall depths as discussed later. For example during one case within the study period, the samplers used 23 bottles to collect samples for a morning rainfall event and only had one collection bottle left for the subsequent afternoon rainfall event. No detailed information is available in the literature providing guidance on programming the sampler to collect a representative number of stormwater samples over the entire inflow hydrographs of natural rainfall events.

Sampler C was activated once 5.7×10^{-5} m^3/s of outflow was detected by an ISCO 750 (Teledyne Technologies, Inc., Thousand Oaks, CA, USA) area velocity flow module (Figure 2a) inserted into the outlet pipe of the skimmer. Sampler C acquired water quality samples from inside the filter grate of the skimmer. Sampler C collected a sample immediately after it was activated by the flow module and continued collecting samples at one-hour intervals until the program was complete, spanning a 24-h period of time and collecting a total of 24, 1 L stormwater samples.

Two other sampler units, Samplers D and E, were positioned within the sediment basin (Figure 1 and Figure 2) to collect water quality samples from a water depth of 0.45 m from the bottom of the basin. The suction heads for the samplers were positioned directly in the middle of Bay 2 (between the first and second baffles, with respect to flow) and in the middle of Bay 3 (between the second and third baffles, with respect to flow). Samplers D and E were connected directly to Sampler C via a custom "Y-cable". This enabled the program for Samplers D and E to collect a sample in sequential order after Sampler C had completed each sampling cycle.

Turbidity and TSS Measurements of Stormwater Runoff Samples

All stormwater samples gathered by the sampling units were transported back to the laboratory for analysis of turbidity and TSS. This allowed for water quality measurements of stormwater runoff at the inflow (just before the weir), within the various bays of the basin, and outflow to be determined. The water quality within the basin was evaluated to allow for determining the settling effect along the flow path from inflow to outflow (as a result of gravity) and any settling enhancement caused by the addition of PAM in the inflow channel and the baffles used within the basin.

Turbidity was measured using a portable HACH 2100Q portable turbidimeter (HACH Company, Loveland, CO, USA). The maximum turbidity reading on this instrument is 1000 NTU. In cases where samples exceeded the maximum turbidity, the test sample was diluted using a 1:2 ratio of low-turbidity deionized water according to instructions given in the "*Sample Dilution*" section of HACH Method 8366 [30]. TSS was determined using vacuum filtration according to the "*Determining Total Suspended Solids*" section of HACH Method 8366 [30].

Soil Samples for Determining Particle-Size Distribution

The particle-size distribution of sediment entrained in sediment-laden stormwater is one of the most critical pieces of information required when assessing sediment basin performance. It is costly and time consuming to determine the particle-size distributions for all inflow and in-basin samples collected during the study for large volumes of stormwater emanating from the contributing drainage area into the sediment basin. Therefore, only the particle-size distributions of raw soils in the construction site and sediment captured in the basin at the conclusion of the study were determined using the standard test method [31]. An assumption was made that the particle-size distributions of sediments captured by the basin should be similar to the particle-size distribution of raw soils collected on-site within the contributing drainage area.

To obtain the particle-size distributions of sediments deposited in the basin, sediment samples were taken at the conclusion of the basin monitoring period in the middle of each bay (Figure 2), with respect to the length of the basin. Three samples were collected in each bay (Bay 1, 2, 3, and 4) with respect to the width of the basin (roughly 2.5 m apart between samples). A total of 12 sediment samples were collected and dried in an oven before the standard test was performed for each sample.

Sediment Basin Survey

An initial, pre-evaluation survey of the sediment basin was performed by ALDOT surveyors immediately after basin construction and prior to the deployment of ISCO sampling units to establish a baseline sediment basin volume. A post-evaluation survey was conducted by the researchers at the end of the monitoring period to determine the volume of sediment captured. Both surveys were performed using Total Stations that provided three-dimensional coordinates of measured points within the basin. To determine the volume difference in the basin due to deposited sediments, a MicroStation (Bentley Systems, Inc., Exton, PA, USA) was used to develop a three-dimensional digital elevation model of the basin. Accuracy of the models was inspected to identify irregular topography of deposited sediment accumulation that did not correlate with visual observations. A retained sediment volume report was then generated by MicroStation, noting the net change in volume that was determined by subtracting the end-volume of the post-evaluation survey from the original volume of the pre-evaluation survey.

SEDIMENT BASIN AT FRANKLIN COUNTY, AL, USA

Study Site

The 12.5 km highway construction project consisted of constructing additional lanes on SR-24 from the east of Red Bay to Dempsey in Franklin County, AL. Figure 2 provides a plan view along with an isometric view of the actual sediment basin monitored as part of this study along with the placement of monitoring equipment. This was the first ALDOT implementation of the 2009 sediment basin design configuration (Figure 1). An equipment installation plan (Figure 2a) was developed prior to construction of the sediment basin. The outflow through the floating skimmer (Figure 2b,c) did not directly discharge into an adjacent stream, but was discharged into an infiltration basin for additional detention time and further water quality protection of the adjacent stream.

The volume of the sediment basin was originally designed to accommodate 512.3 m^3 of stormwater runoff. The bottom length and width of the basin were designed as 23.2 m by 7.0 m, respectively; resulting in a length to width ratio that was 3.3:1, exceeding the minimum 2:1 ratio suggested by the Alabama Handbook [7]. A minor field adjustment during construction added an extra 0.46 m of depth (74.2 m^3 of additional volume of storage) to the basin. This added volume is for sediment storage and dead storage. Thus, the as-built storage volume became 574.5 m^3.

The 2006 design standard requires a minimum basin storage of 126.0 m^3/ha of contributing area draining into the basin [5]. The total contributing watershed area of the basin monitored was 37,272 m^2; therefore, the sediment basin was originally designed and sized using the 2006 design standard. This design limits a sediment basin to have sufficient volume to capture 12.7 mm of effective rainfall per hectare of disturbed area [32].

A 63.5 mm skimmer shown in Figure 1 and Figure 2, which can have a maximum 63.5 mm orifice size, was used as the primary dewatering or outflow control device for the basin. The skimmer was level at 0.46 m above the basin floor. The flow rate from the skimmer was field measured to be 0.0014 m^3/s. The dewatering time of the basin varies with the water level inside the basin. When the basin is full at 1.98 m in depth, the dewatering time is 4.05 days (97.3-h) and discharges a total of 500 m^3 stormwater runoff from the basin into the infiltration basin, excluding dead storage [33].

Based on construction-site soil samples, Applied Polymer Systems (APS) Inc. (Atlanta, GA, USA) recommended using a type 706 B PAM floc block with a reaction or contact time between 40 and 45 s. The recommended dosage rate was 0.0032 to 0.0038 m^3/s of flow per each PAM block placed in a series or in a row.

Data Collection Phases

The total contributing drainage area draining to the basin was constantly evolving during the first four months of the project. Due to the construction progress on a large grading operation (cut) taking place in the upstream area of the basin (Figure 2b), the data collection effort was divided into two phases. Two inflow channels were constructed and used to convey stormwater runoff into the basin. During the first phase (Phase I—15 to 29 November 2011) of monitoring, a single inflow channel (identified as the secondary inflow channel in Figure 2b) was constructed to carry stormwater runoff into the sediment basin. During the second phase (Phase II—30 December 2011 to 16 February 2012), two inflow channels existed as a result of a newly added inflow channel acting as the primary inflow channel for the basin. The ISCO 6712 Sampler B was used to monitor the inflow of stormwater into the basin from the primary inflow channel that was constructed parallel to the constructed road bed, and Sampler A monitored inflow from the secondary inflow channel that conveyed runoff emanated from the hillside as shown in Figure 2.

PERFORMANCE DATA OF THE SEDIMENT BASIN

Phase I Results

The total rainfall depths and the maximum five-minute rainfall intensities for the two events (16 November 2011 and 5 December 2011) during Phase I are shown in Table 1. In addition, five rainfall events resulted in very few inflow samples being collected during Phase I and were not included in the statistical summary provided in Table 1. However the ISCO samplers did collect stormwater samples in Bays 2 and 3 along with the outflow, which allowed us to understand the sediment settling characteristics in the basin as documented by Logan [33]. Even though the total rainfall depths for both events (16 November 2011 and 5 December 2011) were almost the same; peak discharge of stormwater runoff on 16 November 2011 (0.095 m^3/s, Figure 3) was much higher than that on 5 December 2011 (0.035 m^3/s,Figure 3). This was primarily due to the difference in rainfall intensity even though other factors (e.g., grading activity) may also have had an effect. In Phase I of construction, vegetation or ground cover changed over time. The total inflow runoff volume into the basin was 196.5 m^3 for the rain event on 16 November 2011 when the vegetation cover was approximately 10%. The inflow volume was 176.1 m^3 for the rain event on 5 December 2011 after more vegetative growth and ground cover (*i.e.*, approximately 25%) had been established. Based on the depth-storage curve, the maximum water depths in the basin were about 1.02 m and 0.94 m, including 0.46 m of dead storage (74.2 m^3), on 16 November 2011 and 5 December 2011, respectively. Observed inflow turbidity and TSS for the 16 November 2011 rain event had average values of 5855 NTU and 5430 mg/L, respectively. Average observed turbidity and TSS were 1989 NTU and 1305 mg/L for the 5 December 2011 rain event due to smaller rainfall intensity and more vegetative cover despite experiencing almost the same total rainfall depth.

Table 1. Statistical summary of turbidity and total suspended solids (TSS) of inflow stormwater samples.

Date [1]	Rainfall [mm (mm/h)] [2]	No of Data	Turbidity (NTU)				TSS (mg/L)			
			Max	Min	Avg.	STD	Max	Min	Avg.	STD
16 November 2011 [3]	34.3 (56.3)	23	10,656	1030	5855	2582	10,545	790	5430	2689
5 December 2011 [3]	27.7 (27.4)	21	2724	878	1989	446	1950	465	1305	380
17 January 2012 [4]	31.0 (83.3)	28	28,352	3488	9902	6234	26,325	2720	7433	5632
26 January 2012 [4]	18.5 (12.2)	24	785	191	506	149	435	95	275	75
1 February 2012 [4,5]	11.7 (24.4)	24	3688	508	1905	1067	2645	250	1105	745
4 February 2012 [4]	22.4 (15.2)	27	3892	616	1944	914	2315	255	1068	561

Notes: [1] Eight rainfall events with a few inflow samples and two events on 7 and 8 January 2012 in which the samplers were not able to collect inflow samples are not listed in the table above. These rainfall events occurred on 27 November (two events), 28 November, 22 and 26 December in 2011 for Phase I; and on 7, 8, 11, 22 January, and 1 February (evening event) in 2012 for Phase II; [2] total rainfall in mm, and the number inside brackets is the maximum 5-min rainfall intensity in mm/h; [3] rainfall event during the Phase I data collection; [4] rainfall event during the Phase II data collection; [5] rainfall event in the morning of 1 February 2012 (Phase II data collection, there is another event in the afternoon on the same day).

Figure 3 shows that the first measured turbidity (1646 NTU), when outflow began discharging through the skimmer at 9:07 am on 16 November 2011, was much lower than the maximum inflow turbidity of 10,656 NTU (Table 1). This is primarily due to the basin containing inflow runoff collected in the dead storage for approximately 27 min (inflow starting on 8:40 am, Figure 3) prior to the initiation of discharge. Measured turbidity at Bay 2, Bay 3, and outflow, decreased with time to about 240 NTU after 24 h of data collection (Figure 3). The turbidity of 240 NTU was a 98% reduction from the maximum inflow turbidity and a 70% reduction from the first measured turbidity at the outflow. For the 5 December 2011 rain event, the first measured turbidity at the skimmer was only 60 NTU when the outflow started at 3:37 P.M., 15 min after the inflow started at 3:22 P.M. This was because the basin contained less turbid water (after most sediments settled out of suspension) from a previous rainfall event (27–28 November 2011) that diluted the sediment-laden inflow [33]. Measured turbidity of the outflow through the skimmer increased to 1104–1112 NTU from 6:37 P.M. to 8:37 P.M. when sediment-laden stormwater continuously flowed into the basin, then started to decrease at 9:37 P.M. (964 NTU, Figure 3) after inflow stopped, eventually dropping to about 600 NTU at the end of the 24 h data collection period. The turbidity measurement of 600 NTU was a 78% reduction from the maximum inflow turbidity (2724 NTU) and a 46% reduction from the maximum measured turbidity at the outflow (1112 NTU) for the 5 December 2011 rainfall event.

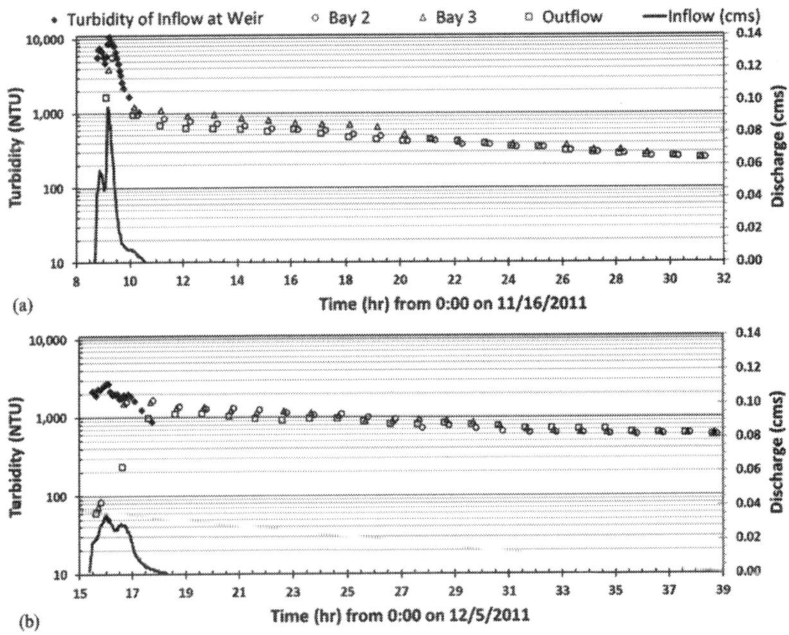

Figure 3. Time-series plots of inflow discharges (cms or m^3/s) and measured turbidity (NTU) from stormwater samples collected at the inflow weir, Bay 2, Bay 3, and outflow of the basin: (**a**) for 16 November 2011 rain event and (**b**) for 15 December 2011 rain event during Phase I.

The difference in performance of the basin for these two rain events was primarily due to the construction contractor applying the correct soil specific PAM used on 16 November 2011 and incorrect PAM used on 5 December 2011. The cause of using incorrect PAM by the contractor is described in Section 5 "Lessons Learned". These variances in performance could also be attributed to different sediment particle sizes flowing into the basin: With more vegetation cover established on 5 December 2011, lower percentage of coarser sediments and higher fraction of finer sediments could have washed into the basin (Figure 4a), resulting in a longer settling time of the smaller particles (requiring additional residence time).

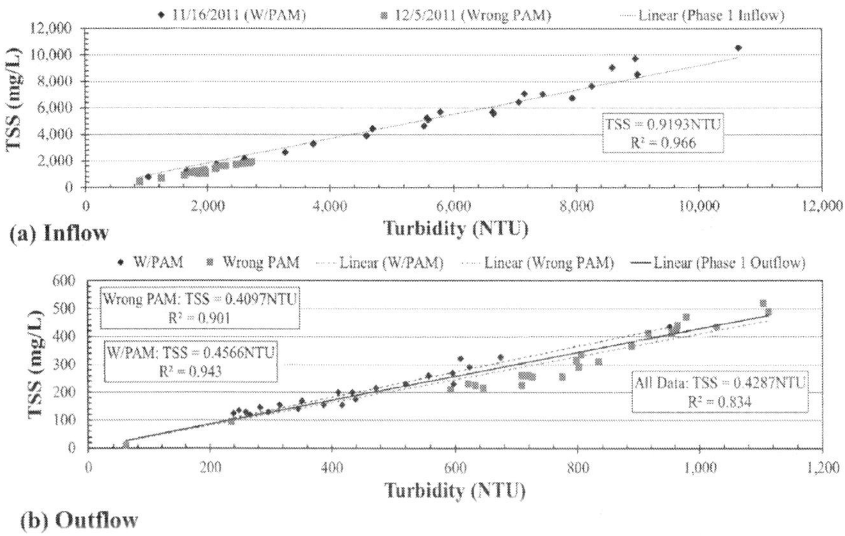

Figure 4. Total suspended solids (TSS) and turbidity correlations for the two data sets in Phase I: Different scales used in (**a**) and (**b**) for x and y axes.

Figure 3 also shows that measured turbidity using stormwater samples, collected at 0.46 m above the basin floor of Bay 2 and Bay 3, were somewhat higher than measured turbidity of outflow collected near the water surface (through the skimmer) in the first 12-h of data collection. After the 12-h data collection, measured turbidity values at Bay 2, Bay 3, and outflow are very similar (Figure 3) despite different depths used for the three stormwater sample collection locations. This suggests that the floating surface skimmer may not always improve discharge water quality when compared to a perforated riser structure—a similar finding from a study in North Carolina by McLaughlin *et al.* [17]. The larger and heavier particles settled first, after a certain retention time (e.g., 12-h for the basin), and only non-settleable, very fine sediments remained suspended in the basin. The skimmer did prevent the discharge of highly turbid water that contained heavy sediments in the first 12-h of the basin dewatering process.

TSS and turbidity correlations were determined separately for the inflow and outflow samples collected during Phase I (Figure 4) and Phase II [33] due to the fact that significantly different correlations exist between the heavier suspended sediments in the inflow and the finer fraction contained in the outflow. The slopes of the linear corrections between TSS and turbidity (NTU) were determined to be 0.9193 (Figure 4a) and 0.4287 (Figure 4b) for inflow and outflow data, respectively. The slopes of the linear relationships between TSS and turbidity for stormwater samples in Franklin County are similar to, but not the same as, the slopes determined from the data of five different sediment detention devices for a study in a North Carolina construction site without distinguishing inflow and outflow samples [13]. Compared to TSS analysis, turbidity is easier and quicker to measure, however the soil-specific relationship of TSS and turbidity has to be established for each monitoring site before one can predict TSS from turbidity [13]. The correlations of TSS and turbidity determined for outflow under W/PAM (16 November 2011) and wrong PAM (5 December 2011) conditions are similar (Figure 4b). This is most likely because large sediment particles (either natural or formed due to flocculation of PAM) had settled before reaching the outflow and finer particles near the water surface were about the same. The TSS data (red squares inFigure 4a) for the inflow on 5 December 2011 indicate that finer or lighter particles were washed into the basin from the site drainage areas as explained above.

Phase II Results

Ten rainfall events occurred during the second phase (Phase II) of the data collection effort, and fifteen 24-h data collection periods of stormwater samples were developed in the basin from 30 December 2011 to 5 February 2012. The data collection in each 24-h period was continuous as shown in Figure 5. The event rainfall depths ranged from 6.6 to 31.0 mm, and the maximum five-minute rainfall intensities for the events ranged from 6.1 to 88.4 mm/h. The rain event on 17 January 2012 had a 31.0 mm rainfall with a maximum five-minute rainfall intensity of 88.4 mm/h. The event on 17 January 2012 also had 15.2 mm of rainfall over a 15-min period (3:30 P.M. to 3:45 P.M.) that produced a large concentration of inflow. A full set of inflow samples was taken to allow for some representation of stormwater runoff entering the basin (Table 1). Measured turbidity for the concentrated inflow on 17 January 2012 ranged from 3488 to 28,352 NTU with an average value of 9902 NTU over the 1.5-h data sampling period, and corresponding TSS ranged from 2720 to 26,325 mg/L (Table 1). These data demonstrate that high intensity rainfall impulses can potentially erode a large amount of sediment from a construction site, generating inflows with very high turbidity and TSS, which creates an additional sediment load for the basin to treat.

The data collected during Phase II were divided into two conditions based on site flow characteristics: (1) "*No PAM*" (four events from 7 to 22 January 2012); and (2) "*W/Limited PAM*" (six events from 26 January to 5 February 2012). Due to the primary inflow channel and weir being improperly installed by the contractor, stormwater flowing into the basin through the primary inflow channel flowed around and under the weir. Under this condition, flow did not come in contact with the floc blocks that were placed on top of the riprap downstream of the weir; therefore, PAM was never introduced to the stormwater runoff during the first condition. After reinstallation of the weir on 24 January 2012, the weir was able to maintain proper function allowing water to flow over the spillway and into the basin. However, the amount of flow occurring over the weir (up to 0.06 m^3/s, Figure 5) resulted in a *limited PAM* treatment condition because the four PAM blocks placed in the inflow channel were only capable of dosing 0.0151 m^3/s (240 gpm) of flow based upon manufacturer's recommendations.

Figure 5 shows example results of Phase II data collection, which includes: Time-series plots of five-minute rainfall intensity (mm/h) and measured turbidity from stormwater samples collected at Bays 2 and 3, and the outflow of the basin, for rain events from 26 to 30 January 2012 (top) and from 1 to 5 February 2012 (bottom). There were three 24-h data collection periods during and after the rainfall event on 26 January 2012, which gave us useful information on sediment settling over an extended period of time. Measured turbidity at the outflow slowly decreased from 905 to 243 NTU over three 24-h monitoring periods with a limited PAM treatment. The maximum turbidity during the event on 26 January 2012 was not very high because of low rainfall intensity, less than 12.2 mm/h (Table 1).

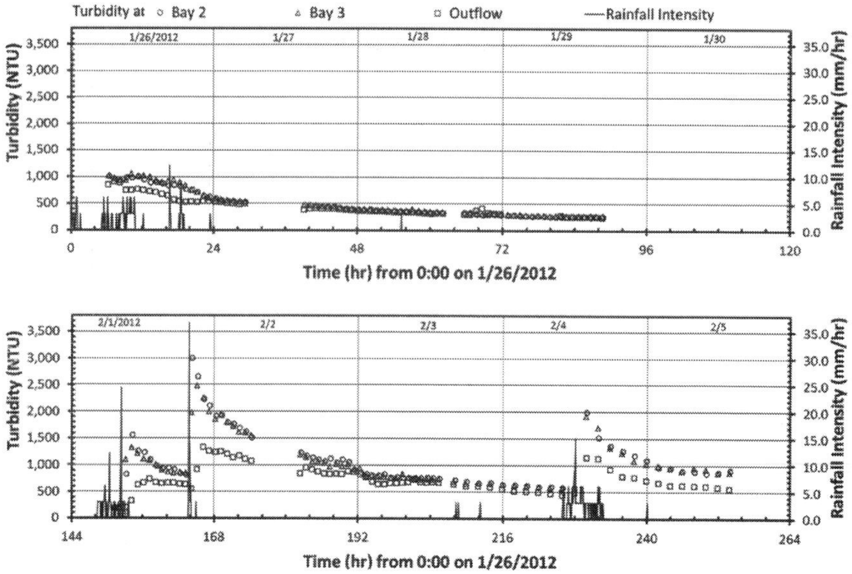

Figure 5. Time-series plots of five-minute rainfall intensity (mm/h) and measured turbidity from stormwater samples collected at the outflow, Bay 2, and Bay 3 of the basin for rain events from 26 January 2012 to 5 February 2012 (Phase II).

There were two rain events on 1 February 2012, from 4:25 A.M. to 9:40 A.M. and from 7:40 P.M. to 8:05 P.M. The morning event on 1 February 2012 had 11.7 mm rainfall with a maximum five-minute intensity of 24.4 mm/h and resulted a maximum turbidity of 3688 NTU at the inflow weir (Table 1). The evening event on 1 February 2012 had only 6.4 mm of total rainfall but had a maximum five-minute intensity of 36.6 mm/h. Measured turbidity at Bay 2 decreased from 1552 to 811 NTU (Figure 5) during, and a few hours following, the morning event. The turbidity jumped to 2996 NTU (Figure 5) due to the evening event on 1 February 2012. Measured turbidity at Bay 2 continued to decrease on 2 February and 3 February 2012 until the rainfall event started on 4 February 2012 that caused a turbidity increase to 1988 NTU. Figure 5 clearly shows the impacts of a subsequent rain event that followed a previous rain event on turbidity in Bay 2, Bay 3, and outflow. The sudden increases in turbidity and TSS [33] just after a subsequent rain event were a result of additional sediment-laden runoff with higher turbidity from the subsequent rain event and possible resuspension of deposited sediment within the basin when additional inflow was introduced.

When the basin was partially filled with runoff and baffle height restricted short circuiting (Figure 2b),Figure 5 clearly shows measured turbidity of collected stormwater samples at 0.46 m above the basin bottom at Bay 2 and Bay 3 could be significantly larger than turbidity of the outflow measured near the water surface. For example, measured turbidity at the outflow followed by the evening event on 1 February 2012 reached a maximum value of 1330 NTU in comparison to the maximum values of 2996 NTU and 2472 NTU near the bottom of Bay 2 and Bay 3, respectively (Figure 5), which illustrates the benefit of using a floating surface skimmer as the primary dewatering device.

The data analysis [33] shows that turbidity and TSS in the basin exponentially decreased with time when there was no inflow entering the basin:

$$NTU\,(t) = NTU\,(t_o)\cdot exp\,(-k_1\cdot t) \tag{1}$$

$$TSS\,(t) = TSS\,(t_o)\cdot exp\,(-k_2\cdot t) \tag{2}$$

where t is the time in hours after the reference time t_o, k_1 and k_2 are decay regression constants (h^{-1}) for turbidity and TSS, respectively. The data analysis indicated that the decay constants of the exponential settling rates ranged from −0.008 to −0.073 h^{-1} [33], which were equivalent to a 17% to 83% reduction in turbidity or TSS per day. During the rainfall period, turbidity and TSS in the basin varied with time depending on several factors: (1) turbidity of inflow; (2) resuspension of previous deposited sediment; and (3) whether the basin was empty or partially filled with stormwater from a previous rainfall event (Figure 2), as illustrated in Figure 5.

Overall Basin Performance

Due to data collection complications in the basin, only two events (16 November 2011 and 5 December 2012) were used to quantify the sediment removal efficiency of the basin during Phase I. The event mean concentrations (EMCs) [34] for TSS (mg/L) and turbidity (NTU) were calculated for inflow and outflow of the basin and reported in Table 2 including the total sediment load (kg). Calculated sediment leaving the basin through outflow was only 25.3 kg, therefore, the efficiency of the basin to remove TSS was 97.9% for the rainfall event on 16 November 2011. TSS load of the outflow on 5 December 2011 was 36.6 kg, which was higher than the outflow sediment load on 16 November 2011, and may be due to suspension of fine sediments deposited in the basin from previous rainfall events. Plus the incorrect PAM being used by the contractor, the efficiency of the basin to remove TSS was 83.7% for the rainfall event on 5 December 2011.

Table 2. Event mean concentration (EMC) for turbidity, TSS, and removal efficiency for two rain events.

Rainfall Event	Inflow (Weir)			Outflow (Skimmer)			Removal Efficiency		
	EMC TSS (mg/L)	EMC Turbidity (NTU)	TSS Load (kg)	EMC TSS (mg/L)	EMC Turbidity (NTU)	TSS Load (kg)	by TSS	by NTU	by Load
16 November 2011	6519.6	6830	1197.7	221.5	478	25.3	96.6%	93.0%	97.9%
5 December 2011	1331.2	2024	224.0	319.6	793	36.6	76.0%	60.8%	83.7%

To quantify total sediments captured by the basin during the study period (all 16 rainfall events), the post-evaluation survey on 26 April 2012 was compared to the pre-evaluation survey of the basin [33]. Total retained sediment volume for the basin was 48.1 m^3 that resulted from sediment-laden runoff generated from rainfall events from 13 September 2011 (when the basin construction was complete) to 26 April 2012. The retained sediment volume occupied 65% of the basin's dead storage (74.2 m^3) and was only 8.4% of the total sediment basin volume (574.5 m^3).

Figure 6 shows average particle-size distributions of raw soils and sediment deposited in each bay of the basin determined from three soil sample analyses. Each processed sample used 1000 g of soil sieved through No. 4 to 200 (particle sizes from 4.76 to 0.074 mm) gradations. The settling velocities of suspended particles in a sediment basin depend on various factors such as particle size or diameter, shape, density, water temperature (viscosity), flow turbulent characteristics, wind wave, use of flocculants, type and amount of flocculants used, *etc.* From the particle-size distribution of raw soils, about 25% of sediments have particle diameters greater than 0.5 mm (coarse sand) with approximate settling velocity of 4.5–6.0 m/min [35], which can settle out of suspension very quickly in the basin. About 47% of sediments are fine or medium-size sand with particle diameters ranging from 0.08 to 0.5 mm that have approximate settling velocity about 0.15–3.0 m/min [35]. With 0.15 m/min of settling velocity, the particle could take less than 15 min to settle from the surface to the bottom under full basin conditions (1.98 m depth). About 15% of sediments were silt, with particle diameters ranging from 0.002 to 0.05 mm that have approximate settling velocities ranging from 0.3 to 0.6 m/h [35]; which would take less than 6.5 h to settle over 1.98 m of depth within the basin. The particle-size distribution (Figure 6) determined using sieve analysis did not report clay faction (diameter <0.0001 mm) in both the raw soil and deposited sediments. However, the settling behavior of sediments in the basin (Figure 3 and Figure 5) indicates that the small percent of clay dominantly controlled and affected turbidity in the sediment basin after 7-h of retention.

The particle-size distributions of deposited sediments indicate that in comparison to other bays and the raw soil, Bay 4 had finer sediments since the curve is above all other distribution curves (Figure 6). Bay 4 had 40% of particles less than 0.1 mm (fine sand). The particle-size distribution in Bay 3 is close to the distribution of raw soil. Bay 1 and Bay 2 contained a greater concentration of larger particles (>0.2 mm), and Bay 1 contained the lowest percentage of smaller particles with diameters less than 0.1 mm.

In the post-evaluation survey [33] we collected the coir baffle materials: Three 0.36 m^2 (0.6 by 0.6 m) samples were collected from each baffle. The clean coir baffle used in the basin was measured to be 975 g/m^2. Collected baffle samples between Bay 1 and 2, Bay 2 and 3, Bay 3 and 4 were weighted as 3263; 2005; and 1787 g/m^2; respectively [33]. Therefore, these baffles did capture sediments moving towards the outflow, and baffle densities indirectly indicate sediments from Bay 1 to Bay 4 decreased in particle size and density (became finer).

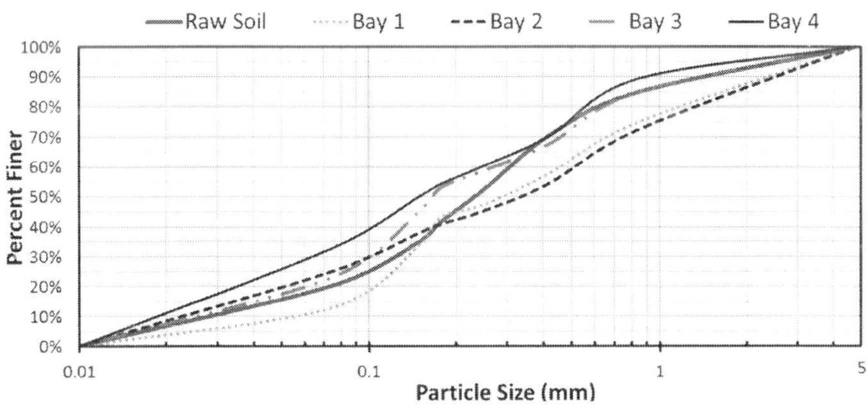

Figure 6. Particle-size distributions of raw soil at the construction site of the basin, deposited sediments in Bay 1, Bay 2, Bay 3, and Bay 4 of the basin at the end of data collection.

LESSONS LEARNED

When lessons learned are summarized and shared, it is beneficial to the professional community and for future studies monitoring sediment basin performance. The lessons learned are explained below as follows:

- The monitoring team needs to provide clear instructions to the construction contractor on where monitoring equipment will be located at the sediment basin. Figure 2a shows a schematic diagram (using the original design drawing as background) of the sediment basin monitored in the study, providing the approximate locations of the monitoring equipment. Figure 2a was developed for the contractor to appropriately construct the basin (*i.e.*, weir installation, baffle placement, site preparation for sampler's installation, *etc.*) for the researchers to install all monitoring equipment immediately after the construction of the sediment basin is complete. A similar diagram was not prepared for a different sediment basin in the same construction project, which was supposed to be monitored. Wet weather conditions prevented the contractor and researchers from properly installing the data collection equipment since site preparation instructions were not provided in advance to the contractor. This resulted in the site not being properly prepared when the basin was originally constructed and the researchers not being able to install monitoring equipment in a timely fashion. Therefore, this particular basin was not monitored and data was never collected resulting in a missed opportunity.
- The monitoring team and the basin design standards need to provide appropriate instructions to the contractor on the proper use, dosage, and placement of PAM floc blocks. For the 2009 sediment basin design used on the ALDOT construction project, the proper placement and number of PAM blocks are important to ensure that sediment-laden stormwater inflow is properly dosed to promote flocculation of suspended sediment and deposition within the basin. Figure 1 is a part of the project drawing for the construction project that shows the placement of four PAM floc blocks in the inflow channel downstream of the rock ditch check and weir structure. The drawing also includes a special project note (not shown in Figure 1) stating:
- "*4. Payment for polyacrylamide block will be paid under item 655-W. Four (4) blocks shall be placed at the inlet end of the basin. Block condition shall be monitored and blocks shall be replaced when they have degraded to the point they no longer appear to be effective.*"

- The contractor followed the drawing and the special project note verbatim and installed four PAM floc blocks after the inflow weir without considering the actual amount of flow expected from the contributing drainage area.
- The monitoring team needs to provide the contractor with specific instructions on using the correct type of PAM that is decided based upon laboratory testing of the local site soil properties. Appropriate PAM floc blocks were purchased by the contractor from APS after the soil samples from the construction site were tested by the manufacturer. After the 16 November 2011 rain event, the contractor cleared a large amount of small #4 aggregate from the rock ditch check that had washed into the weir and accidently removed the correct PAM. The contractor then placed four new PAM floc blocks in the channel downstream of the weir, but during a visit to the basin following the rain event on 5 December 2011, the researcher discovered that those PAM blocks were not the soil prescribed type of PAM (possibly left over from a previous project). Therefore, during the Phase I data collection effort, two conditions were observed: (1) correct PAM placement and (2) incorrect (*i.e.*, wrong type) PAM placement in the inflow channel. These two conditions are referred to as "***w/PAM***" and "***w/wrong PAM***", respectively, when data collection results are discussed.
- If a weir is used for data collection, clear instructions need to be provided to the contractor on the proper installation. The runoff should not be allowed to flow around and under the weir, which will result in inaccurate flow rate measurements. As a result of improper construction and installation of the weir at the primary inflow channel by the contractor (when the monitoring team was not on-site), no or a few inflow samples were collected for three rainfall events (on 7, 8, and 11 January 2012 in Table 1) during Phase II data collection. After the weir was appropriately reinstalled on 24 January 2012, only four sets of complete inflow data were collected for Phase II, which are listed inTable 1.
- The baffle height requirement should be clearly stated (e.g., in notes) and not just shown in the design drawing for the construction and installation of a sediment basin. The purpose of the baffles are to help slow inflow velocities and distribute the flow across the entire width of the basin, thereby creating favorable conditions for sedimentation [7] to occur under non-turbulent flow conditions as shown in Figure 2b. The problem observed during data collection, as shown in Figure 2c, was that when runoff from a rain event completely filled the basin, the water level within the basin overtopped the baffles, creating a fully mixed condition within the

basin, disabling the primary function of the baffles. The runoff from the rainfall event on 26 January 2012 (Figure 5 top) was fully contained in the basin and the water level was lower than the baffle height, therefore, outflow turbidity was low. However, the rainfall events on 1 February 2012 and 4 February 2012 (Figure 5bottom) created overflow through the spillway and the water level was higher than the baffle height, therefore, outflow turbidity was much higher. In the case of the basin monitored, the height of the baffles was 1.22 m when the contractor used available coir baffle materials, whereas the full depth of the sediment basin was 1.98 m. For the baffles to function properly, as designed, it is imperative that the height of the baffles be greater than the maximum potential water level within the sediment basin, just prior to discharge over the emergency spillway.

Challenges of Data Collection and Basin Operations

There were also various challenges associated with the data collection efforts and basin operations for an ongoing construction project, which are summarized below:

- There were six rainfall events that occurred during Phase I of the construction effort, and complete data collections were developed for two of these six rainfall events (16 November 2011 and 5 December 2011, Figure 3). Freezing temperatures and redirecting the runoff as construction progressed, prevented complete data collection for the other four events.
- Due to the large concentrated inflow created on 17 January 2012, damage occurred to the primary inflow channel and weir, rendering the weir useless for future data collection until properly repaired.
- To ensure adequate performance, it is important that the basin is sized appropriately to capture and detain stormwater runoff for the maximum potential drainage area contributing runoff into the basin during the life of the project.
- Resuspension of deposited sediments from previous rainfall events resulted in higher turbidity in the basin and reduced the efficiency of the basin. The period with no rain after 26 January 2012 allowed sediments to continuously settle (Figure 5) but three rain events from 1 to 4 February 2012 resulted in complex and dynamic variations of turbidity and TSS in the basin under a limited PAM condition.
- The turbidity of sediment-laden inflow from a construction site actually depends less on the total amount of rainfall from an event (e.g., 2-year, 24-h design storm) but more on rainfall intensity. For example, the rainfall event on 17 January 2012 had a total rainfall

depth of 31.0 mm with a maximum five-minute rainfall intensity of 88.4 mm/h that generated inflow with maximum measured turbidity and TSS of 28,352 NTU and 26,325 mg/L, respectively. The total rainfall depth for the event was much less than 124.5 mm rainfall for a 2-year, 24-h storm in Franklin County, AL, but the maximum five-minute rainfall intensities for the 2 year and 10 year, 24 h design storms are only 11.2 mm/h and 16.3 mm/h, respectively, using Type II NRCS rainfall distribution [36]. A basin designed using a 2-year, 24-h storm would have adequate storage volume to hold runoff generated from the event, but whether the basin can sufficiently handle high turbidity generated from other high intensity rainfall impulses is unknown and needs to be further studied.

- It is difficult and challenging to dose sediment-laden stormwater inflow with the exact amount of flocculating agent across all rainfall events since rainfall is a stochastic variable. This can be considered as a challenge for specification writing by agencies and basin operations by contractors. This will be discussed in detail in the next section.

DISCUSSION AND RECOMMENDATIONS

The current USEPA's National Pollutant Discharge Elimination System (NPDES) Construction General Permit provides sediment basin design requirements in Section 2.1.3.2-a-i to "Provide storage for either (1) the calculated volume of runoff from a 2-year, 24-h storm; or (2) 3600 cubic feet per acre drained" [37]. The 2009 Alabama Handbook [7] requires a minimum volume of a sediment basin to be 3600 cubic feet per acre (*i.e.*, 251.9 m^3/ha) of the contributing drainage area. Observed inflow volumes to the study basin for four events occurred on 16 November 2011, 5 December 2011, 1 February 2012, and 4 February 2012 were 195.8; 176.1; 620.8; and 805.8 m^3 [33], respectively. Using 251.9 m^3/ha design standard the basin associated with 3.73 ha watershed would provide 938.9 m^3 of storage, which would be sufficient to hold the observed volumes produced by rain events during both Phase I and Phase II data collections. The 251.9 m^3/ha design standard is still insufficient for the basin to detain the runoff volume of a 2-year, 24-h storm event (2749.5 m^3 determined from a rainfall-runoff modeling); therefore, the basin is significantly undersized with respect to handling 2-year, 24-h event runoff volumes.

USEPA [4] estimated the storage volume of a sediment basin using a weighted runoff coefficient based on hydrologic soil group distributions

and the 2-year, 24-h storm depth for a selected location in each state. USEPA [4] calculated an advanced-treatment-system storage requirement for Alabama to be 825.5 m^3/ha using the 2-year, 24-h rainfall depth in Montgomery, which is 3.3 times larger than the current minimum storage requirement in Alabama. An under-designed sediment basin (from a volumetric standpoint) more frequently allowed highly turbid sediment-laden runoff to directly flow over the emergency spillway to a downstream receiving water body. During data collection, five rainfall events occurred on 11, 17, 22 January 2012, 1 and 4 February 2012 resulted in overflow through the spillway of the basin into the infiltration basin. The rainfall events on 17 January 2012 and 1 February 2012 filled the basin up from a zero-outflow point to generate overflow, while the other three events resulted in overflow when the basin was partially full of sediment-laden stormwater runoff from a previous rainfall event (water level was above the 0.46 m dead storage).
In addition, the proper placement of PAM floc blocks is crucial to ensure that sediment-laden stormwater inflow is properly dosed to promote flocculation of suspended sediment and deposition within the basin. Based on the APS's dosage recommendation, four PAM floc blocks placed in the inflow channel downstream of the weir by the contractor would handle a maximum flow of 0.0126 to 0.0151 m^3/s. Figure 7 presents the observed inflow hydrographs for several rainfall events from the Phase I and Phase II data collection, including average inflow rates of 2-year, 24-h and 10-year, 24-h design storms for the basin studied in Franklin County, AL. The 2009 Alabama Handbook [7] suggests designing sediment basins using a 2-year, 24-h rainfall and spillway using a 10-year, 24-h rainfall. Observed inflow rates resulted from these five events exceeded the effective flow rate limit (0.0151 m^3/s) of the four PAM floc blocks that were in place in both inflow channels (Figure 4) during both phases of data collection. Runoff hydrographs of design storms were modeled using NRCS 24-h rainfall distribution (Type II for Franklin County, AL, USA), hydrologic soil group C for newly graded area (Curve Number CN = 91), and a NRCS dimensionless unit hydrograph. The 2-year and 10-year, 24-h rainfall depths are 99.1 mm and 142.2 mm in Franklin County, AL, USA; and produce 2749 m^3 and 4348 m^3 of storm runoff with peak discharges of 0.886 m^3/s and 1.367 m^3/s, respectively. To accommodate or dose for peak discharges, 234 and 360, 706B PAM floc blocks would be required, which seems infeasible. Average inflow rates were calculated and plotted on Figure 7 for the 2-year and 10-year, 24-h design storms; and would require 10 and 15, 706B PAM floc blocks, respectively. For both the 2-year and 10-year, 24-h design storms, there are about 21 h or 87.1% of the runoff hydrograph having discharges less than the corresponding

average inflow rates, respectively, which account for about 32.4% of the total runoff volume. Figure 7shows that it is difficult and challenging to dose stormwater runoff with the exact amount of PAM floc blocks for inflows from all rainfall events that a study area could experience during the design life because rainfall is a stochastic variable.

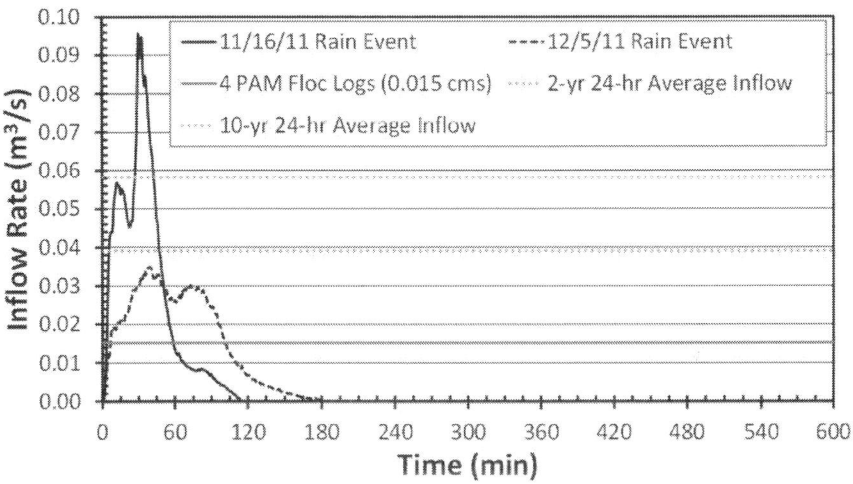

Figure 7. Observed inflow hydrographs for several rainfall events during the Phase I (**top**) and Phase II (**bottom**) data collection including effective flow rate using four floc blocks and average inflow rates for 2-year and 10-year 24-h design storms, where cms is m^3/s.

SUMMARY AND CONCLUSIONS

A field-scale data collection plan to monitor and evaluate sediment basin performance was developed and implemented using automatic stormwater samplers, flow modules, a rain gauge, and inflow weirs. As stormwater management and sediment control practices are employed to protect downstream water resources, the sediment basin on a construction project in Franklin County, AL was monitored from 15 November 2011 to 6 February 2012 over 16 rainfall events when inflow runoff was generated. Through completing the study, the following summary and conclusions have been developed:

- The sediment basin did effectively remove sediments during the early stages of construction when the basin's influent most likely contained a relatively large percentage of large-size sediment particles. The sediment basin removed 97.9% and 83.7% of sediments generated by rainfall events on 16 November 2011 and 5 December 2011, respectively.

- High-intensity rainfall impulses generated high turbidity inflows from the construction site, re-suspended previously settled sediments within the basin, and suddenly increased in-basin turbidity, which could be several times higher than turbidity of water already in the basin.
- The lessons learned and the challenges of the field data collection to evaluate the performance of the sediment basin were listed and summarized, which will benefit the professional community for planning future studies on sediment basin monitoring efforts.

Using 251.9 m^3/ha design standard, the basin associated with 3.73 ha watershed would provide 938.9 m^3 of storage, which would be sufficient to hold the observed runoff volumes produced by rain events during both Phase I and Phase II data collection efforts. Further research is necessary to determine whether the basin should be designed and sized using a 2-year 24-h storm. It is recommended that the height of the baffles match or exceed the full depth of the sediment basin and not be installed below the minimum elevation of the emergency spillway, preventing stormwater from overtopping them and creating a fully mixed condition. Further research is necessary to determine where and how PAM floc blocks can be effectively and practically introduced in the inflow channel system to achieve adequate dosage, agitation, and contact time.

ACKNOWLEDGMENT

This paper is based on a study sponsored by ALDOT, the authors gratefully acknowledge this financial support. The findings, opinions, and conclusions expressed in this paper are those of the authors and do not necessarily reflect the view of the sponsor. The authors thank four anonymous reviewers and Michael A. Perez at Auburn University who provided constructive comments and suggestions that helped authors to improve the quality of the manuscript.

AUTHOR CONTRIBUTIONS

Xing Fang and Wesley Zech analyzed the data and contributed significantly in writing and revising the manuscript. Christopher Logan contributed significantly in collecting the field data for the study, performed initial data analysis, and authored portions of the manuscript.

REFERENCES

1. United States Environmental Protection Agency (USEPA). *Stormwater Phase II Finale Rule—Construction Site Runoff Control Measure*; EPA Federal Register: Washington, DC, USA, 2005; Volume 833-F-00-008.
2. United States Environmental Protection Agency (USEPA). *Effluent Limitations Guidelines and Standards for the Construction and Development Point Source Category*; USEPA: Washington, DC, USA, 2013.
3. National Pollutant Discharge Elimination System Permit—General Perimt. Available online: http://adem.alabama.gov/programs/water/waterforms/ALR10CGP.pdf (accessed on 29 June 2015).
4. United States Environmental Protection Agency (USEPA). *Development Document for Final Effluent Guidelines and Standards for the Construction & Development Category*; USEPA, Office of Water: Washington, DC, USA, 2009.
5. Alabama Soil and Water Conservation Committee (ASWCC). *Alabama Handbook for Erosion Control, Sediment Control, and Stormwater Management on Construction Site and Urban Areas*; ASWCC: Montgomery, AL, USA, 2006; Volume 1, p. 374.
6. Jarrett, A.R. Design of perforated risers to control dewatering of sedimentation basins. *Appl. Eng. Agric.* **1993**, *9*, 37–42.
7. Alabama Soil and Water Conservation Committee (ASWCC). *Alabama Handbook for Erosion Control, Sediment Control, and Stormwater Management on Construction Site and Urban Areas*; ASWCC: Montgomery, AL, USA, 2009.
8. Millen, J.A.; Jarrett, A.R.; Faircloth, J.W. Experimental evaluation of sedimentation basin performance for alternative dewatering systems. *Trans. Am. Soc. Agric. Eng.* **1997**, *40*, 1087–1095.
9. Rauhofer, J.; Jarrett, A.R.; Shannon, R.D. Effectiveness of sedimentation basins that do not totally impound a runoff event. *Trans. Am. Soc. Agric. Eng.* **2001**, *44*, 813–818.

10. Bidelspach, D.A.; Jarrett, A.R.; Vaughan, B.T. Influence of increasing the delay time between the inflow and outflow hydrographs of a sediment basin. *Trans. Am. Soc. Agric. Eng.* **2004**, *47*, 439–444.
11. Thaxton, C.S.; Calantoni, J.; McLaughlin, R.A. Hydrodynamic assessment of various types of baffles in a sediment retention pond. *Trans. Am. Soc. Agric. Eng.* **2004**, *47*, 741–749.
12. Thaxton, C.S.; McLaughlin, R.A. Sediment capture effectiveness of various baffle types in a sediment retention pond. *Trans. Am. Soc. Agric. Eng.* **2005**, *48*, 1795–1802.
13. McCaleb, M.M.; McLaughlin, R.A. Sediment trapping by five different sediment detention devices on construction sites. *Trans. ASABE* **2008**, *51*, 1613–1621.
14. Kang, J.; King, S.E.; McLaughlin, R.A. Impacts of flocculation on sediment basin performance and design. *Trans. Am. Soc. Agric. Eng.* **2014**, *57*, 1099–1107.
15. North Carolina Department of Transportation (NCDOT). Sediment traps and barriers. *North Carolina Erosion and Sediment Control Planning and Design Manual*; NCDOT: Raleigh, NC, USA, 2013; p. 58.
16. Hossain, M.A.; Alam, M.; Yonge, D.R.; Dutta, P. Efficiency and flow regime of a highway stormwater detention pond in washington, USA. *Water Air Soil Pollut.* **2005**, *164*, 79–89.
17. McLaughlin, R.A.; Hayes, S.A.; Clinton, D.L.; McCaleb, M.M.; Jennings, G.D. Water quality improvements using modified sediment control systems on construction sites. *Trans. ASABE* **2009**,*52*, 1859–1867.
18. Bidelspach, D.A.; Jarrett, A.R. Electro-Mechanical outlet flow control device delays sediment basin dewatering. *Appl. Eng. Agric.* **2004**, *20*, 759–763.
19. Bhardwaj, A.K.; McLaughlin, R.A. Simple polyacrylamide dosing systems for turbidity reduction in stilling basins. *Trans. ASABE* **2008**, *51*, 1653–1662.
20. Haan, C.T.; Barfield, B.J.; Hayes, J.C. *Design Hydrology and Sedimentology for Small Catchments*; Academic Press: New York, NY, USA, 1994.
21. Bentzen, T.R.; Larsen, T.; Rasmussen, M.R. Predictions of resuspension of highway detention pond deposits in interrain event periods due to wind-induced currents and waves. *J. Environ. Eng.* **2009**, *135*, 1286–1293.
22. Barfield, B.J.; Clar, M. *Development of New Design Criteria for Sediment Traps and Basins*; Maryland Resource Administration: Annapolis, MD, USA, 1985.

23. Line, D.E.; White, N.M. Efficiencies of temporary sediment traps on two North Carolina construction sites. *Trans. Am. Soc. Agric. Eng.* **2001**, *44*, 1207–1215.
24. Comings, K.J.; Booth, D.B.; Horner, R.R. Storm water pollutant removal by two wet ponds in bellevue, Washington. *J. Environ. Eng.* **2000**, *126*, 321–330.
25. Zhou, S.; McCorquodale, J.A. Modeling of rectangular settling tanks. *J. Hydraul. Eng.* **1992**, *118*, 1391–1405.
26. Pitt, R.E.; Clark, S.E.; Lake, D.W. *Construction Site Erosion and Sediment Controls: Planning, Design, and Performance*; DEStech Publications, Inc.: Lancaster, PA, USA, 2007; p. 381.
27. Madaras, J.S.; Jarrett, A.R. Spatial and temporal distribution of sediment concentration and particle size distribution in a field scale sedimentation basin. *Trans. Am. Soc. Agric. Eng.* **2000**,*43*, 897–902.
28. Bhardwaj, A.K.; McLaughlin, R.A.; Babcock, D.L. Energy dissipation and chemical treatment to improve stilling basin performance. *Trans. ASABE* **2008**, *51*, 1645–1652.
29. Zech, W.C.; Logan, C.P.; Fang, X. State-of-the-practice: Evaluation of sediment basin design, construction, maintenance, and inspection procedures. *ASCE Pract. Period. Struct. Des. Constr.* **2014**, *19*, 04014006.
30. Hach Company. Determining the Relationship between Turbidity and Total Suspended Solids. In*Hach Method 8366 Analytical Procedures*; Hach Company: Loveland, CO, USA, 2011.
31. American Society of Testing and Materials (ASTM). *ASTM d6913-04(2009)e1: Standard Test Methods for Particle-size Distribution (Gradation) of Soils Using Sieve Analysis*; American Society of Testing and Materials (ASTM) International: West Conshohocken, PA, USA, 2009.
32. North Carolina Department of Transportation (NCDOT). Sediment traps and barriers. *Practice Standards and Specifications*; NCDOT: Raleigh, NC, USA, 2006.
33. Logan, C.P. Assessing Performance Characteristics of Sediment Basins Constructed in Franklin County of Alabama. Master's Thesis, Auburn University, Auburn, AL, USA, 2012.
34. Wanielista, M.P.; Yousef, Y.A. *Stormwater Management*; John Wiley & Sons, Inc.: New York, NY, USA, 1993.
35. U.S. Department of Agriculture (USDA). *Soil Survey of Porter County, Indiana*; USDA Soil Conservation Service: Washington, DC, USA, 1981.
36. Viessman, W.; Lewis, G.L. *Introduction to Hydrology*, 5th ed.; Pearson Education: Upper Saddle River, NJ, USA, 2003; p. 612.

37. U.S. Environmental Protection Agency (USEPA). *National Pollutant Discharge Elimination System General Permit for Discharges from Construction Activities*; USEPA: Washington, DC, USA, 2012.

CITATION

Xing Fang, Wesley C. Zech 1 and Christopher P. Logan, Stormwater Field Evaluation and Its Challenges of a Sediment Basin with Skimmer and Baffles at a Highway Construction Site, doi:10.3390/w7073407.

CHAPTER 3

Traffic Recovery Time Estimation Under Different Flow Regimes in Traffic Simulation

Mansoureh Jeihani, Petronella James, Anthony A. Saka, Anam Ardeshiri

Department of Transportation and Urban Infrastructure Studies, Morgan State University, Baltimore, MD 21251, USA

ABSTRACT

Incident occurrence and recovery are critical to the smooth and efficient operations of freeways. Although many studies have been performed on incident detection, clearance, and management, travelers and traffic managers are unable to accurately predict the length of time required for full traffic recovery after an incident occurs. This is because there are no practical studies available to estimate post-incident recovery time. This paper estimates post-incident traffic recovery time along an urban freeway using traffic simulation and compares the simulation results with shockwave theory calculations. The simulation model is calibrated and validated using a freeway segment in Baltimore, MD. The model explores different flow regimes (traffic intensity) and incident duration for different incident severity, and their effects on recovery time. A total of 726 simulations are completed using VISSIM software. Finally, the impact of congestion and incident delay on the highway network is quantified by a regression formula to predict traffic recovery time. The developed regression model predicts post-incident traffic recovery time based on traffic intensity, incident duration, and incident severity (ratio of lanes closure). In addition, three regression models are developed for different flow regimes of near-capacity, moderate, and low-traffic intensity. The model is validated by collected field data on two different urban freeways.

INTRODUCTION

Freeway congestion is a major problem in many urban areas. Congestion on freeways is classified to recurring and non-recurring. Recurring congestion is from normal peak-hour travel. Non-recurring congestion is from random and unpredictable incidents and events that impede the flow of traffic, such as lane blockage from accidents, disabled vehicles, or natural phenomena. These non-recurring incidents can make large delays that contribute significantly to the total congestion experienced by travelers. Delays are influenced by the nature and frequency of incidents and the traffic intensity before the incident. Accurate estimations of congestion delay and its components are important for effective traffic management. Traffic management decisions will be largely influenced by the nature and type of congestion experienced. Traffic management strategies should be emphasized if the accrued delay is largely from recurrent congestion, and the incident management strategies should be applied if the delay is largely incident related (Skabardonis and Geroliminis 2004).

A literature search is conducted to find past researches related to incident delay estimation and recovery time. The obtained information indicates that most of the available studies utilized the analytical model of queuing analysis (Garib et al., 1997;Giuliano, 1989; Lindley, 1987; Morales, 1986; Olmstead, 1999; Sullivan, 1997) and shockwave analysis theory (Hadi et al., 2007; Knoop, 2010). While these methodologies remain popular, others have concluded that these approaches underestimate the actual queue dissipation time and, ultimately, the full system recovery time (Chien and Chowdhury, 2000; Li et al., 2006). Although these analytical models can reasonably estimate the average delay, they seriously underestimate the standard deviation of delay and the expected total delay in the dynamic traffic networks.

Delay is one of the most important indicators to measure the impacts of incidents. Several methods (queuing and shockwave) are available in the literature for incident-induced delay estimation on freeway networks. The deterministic queuing model (DQM) is one of the most widely used methods and also supported by the Highway Capacity Manual (TRB, 2010).

DQM and shockwave theory are often used to evaluate the characteristics of queue formation and dissipation. DQM is based on assumptions regarding arrival patterns, departure characteristics, and queue disciplines.

The queue discipline that most readily assumed for traffic-oriented queues is the first-in, first-out (FIFO).

A shockwave means a discontinuity of flow or density and occurs when cars change speed abruptly. A sudden reduction of the freeway capacity creates backups and queuing, and results in the shockwave effect. The sudden reduction of capacity results from either recurring or non-recurring congestion. The bottleneck results in speed reduction, and the point at which this change occurs can be noted by the brake lights on the vehicles.

According to Skarbardonis and Geroliminis (2004), simulation models can be applied to analyze incident impacts without simplifying assumptions which is required by analytical techniques. Furthermore, most previous studies have only estimated the queue dissipation time, and had no standard formulation for full traffic recovery time (TRT) estimation. Therefore, traffic managers in different areas have postulated that post-incident TRT exceeds the actual duration of an incident by a fixed factor. For example, this factor is postulated to be four and ten in Maryland and California, respectively (Chang et al., 2006). While that idea is clearly refutable because the recovery time is a function of the prevailing traffic intensity, it does have some element of truth regarding the relatively longer period of traffic recovery and the actual duration of the incident. In this study, TRT is defined as the time when post-incident traffic flow has returned to pre-incident conditions.

It is usually difficult to accurately predict the length of time required for full traffic recovery after an incident. The probabilistic nature of most non-recurring incidents makes it difficult to collect accurate empirical data to establish a mathematical relationship between incident duration and TRT for different flow regimes or traffic intensity values. The duration of most non-recurring incidents is usually unknown because of one's inability to determine the exact time of occurrence. Microscopic simulation allows for generation of pseudo-incidents for a variety of traffic-flow scenarios. These pseudo-incidents can facilitate a controlled study on the ramifications of delay to highway incident response.

A typical time-density-speed graph of incidents is presented in Fig. 1 to show the difference between queue dissipation and full traffic recovery. The upper line segment in the graph represents the density curve in vehicle per mile (vpm), while the downward slope of the line represents the queue discharge during the traffic stabilization period prior to the onset of full TRT, or pre-incident conditions. The lower line represents the speed curve in miles per hour (mph). The first section is the pre-incident normal condition. The incident begins at T1 and ends at T2. Queue dissipation starts at T2 and ends at T3. Full traffic recovery happens at T4. The time between T1 and T2 is the incident duration when an incident happens, lanes are closed until the incident is cleared and lanes would be re-opened. During the incident, density increases and speed decreases since one or more lanes of the freeway are blocked. After the incident ends, recovery begins and traffic dissipates. Although the queue is dissipated at T3, the traffic is not stabilized. Full incident recovery is achieved when pre-incident conditions are observed, after queue dissipation at T4. The authors considered both speed and density for traffic recovery. Density is a more accurate indicator for traffic congestion along freeways, as freeways can be heavily congested even at free flow speeds.

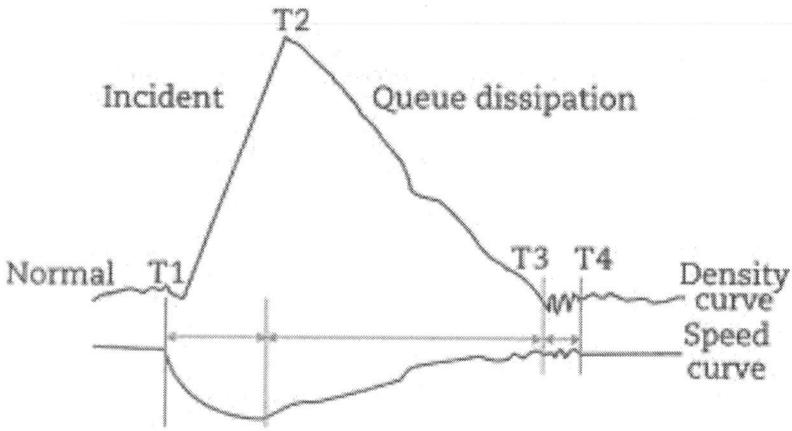

Figure 1. Typical time-density-speed graph of incidents and traffic recovery.

Computer simulation models have become increasingly important in the analysis, design, and management of transportation/traffic infrastructure and operations. This is particularly true for delay impact, delay analysis, incident detection, and incident management, which form the complex and frequently changing traffic conditions. Since it is expensive and difficult to analyze such situations through empirical methods (due to the large amount of data required), simulation models are often used. In most cases, only limited, if any, field tests are feasible, because of prohibitively high costs and lack of public acceptance (Toledo and Koutsopoulos, 2004). Boyles and Waller (2007)demonstrated that the incident impact is underestimated by 20 percent to 50 percent, due to the uncertainty in predicting incident duration. They developed a stochastic delay prediction model to mitigate the underestimation of incident impact. They also employed a mesoscopic simulator with Monte Carlo sampling to study incident delay for different demand profiles. Simulation models also help engineers evaluate alternative transportation strategies and predict outcomes of possible improvements to the transportation systems. The most popular microscopic traffic simulation programs are CORSIM, INTEGRATION, WATSIM, TRANSIMS, MITSIM, PARAMICS, VISSIM, and AIMSUN (Zhang and Owen, 2004). There are several contributions in the literature on calibration/validation of simulation models to match field conditions (Dowling et al., 2004;Jha et al., 2004; Ni, 2004; Toledo and Koutsopoulos, 2004).

The objective of this study is to present a simulation-based methodology for incident TRT estimation. The present research uses traffic simulation to explore the relationship among incident recovery time, traffic intensity, incident severity, and incident duration, and to compare the simulated data with the result of analytical shockwave traffic recovery estimation model. The relationship is demonstrated using the results from the VISSIM 4.30 traffic simulation model (PTV, 2000). The estimation formula is developed using regression models.

Traffic managers can use the developed formulas to calculate the full TRT based on traffic intensity, incident severity, and incident duration. The research is expected to serve as a valuable guide for incident managers and decision makers as they assess the ramifications of delayed response to highway incidents and develop improved incident-management methods. This research enhances management agencies' ability to quantify the impact of congestion and delay on the freeway network. The improved congestion management also increases the reliability of traffic prediction in advanced transportation information systems and, ultimately, the social welfare of commuters and drivers.

MATERIALS AND METHODS

This section discusses the research process and highlights the simulation procedure, determination of the TRT from the simulation outputs, and the shockwave delay calculations.

The authors developed a traffic simulation model to generate traffic intensity and incident duration for different incident severity (lane closure) scenarios. Then, the TRT for each scenario using the developed simulation model was derived. The queue dissipation time (QDT) for each scenario was also calculated using shockwave theory. The results of the models were compared. Finally, the authors developed a regression model using the simulation results to estimate TRT for each lane closure scenario (incident severity), traffic intensity, and incident duration.

Traffic simulation

The authors developed a traffic simulation model for a typical three-lane unidirectional urban freeway. The model generated different incident scenarios of various traffic intensity and duration for different lane closures. Driver behavior parameters were adjusted and included the "look-behind distance" and the "lane-changing" parameters. This eliminated collisions and mini-queues had reduced the overestimation of the queue dissipation time. The VISSIM simulation model was previously calibrated and validated in an earlier study (Saka et al., 2008). A modified Chi-square test known as GEH was employed to validate the model. The calibration and validation was performed on the I-83 freeway corridor known locally as the JFX in Baltimore, MD, by iteratively comparing the model output to the observed traffic performance. Adjustments were made as needed to reasonably replicate the observed condition. Fig. 2 presents the JFX corridor and Table 1presents a summary of the calibration/validation process, which is based on Oketch and Carrick's criteria (Oketch and Carrick, 2005). Parameters used in the simulator were based on the Wiedemann (1974, 1991) approach or software default values, or obtained by measurement of traffic on the freeway.

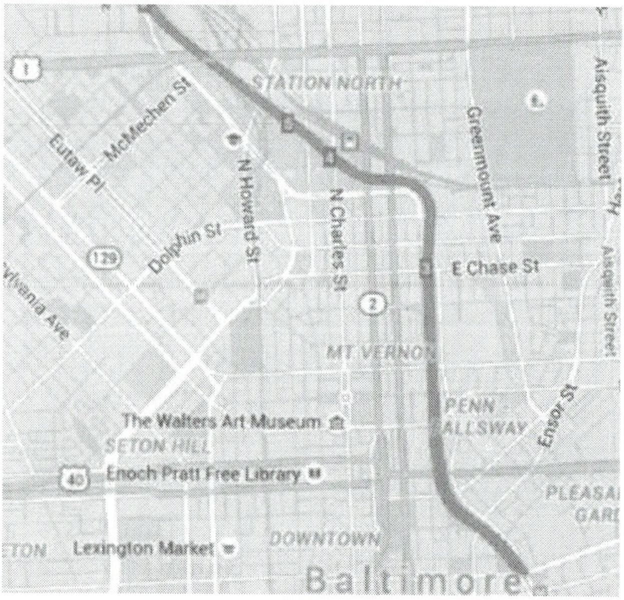

Figure 2. JFX Corridor.

Table 1. Observed versus simulated throughputs in study area.

JFX segment southbound	Observed volume o(vph)	Simulated volume Range e (vph)	GEH $\left(\sqrt{\frac{(o-e)^2}{0.5(o+e)}}\right)$	Validation criteria (GEH < 5) met?
Between Exit 5 and Exit 4	8075	7595–7879	2.2	Yes
Between Exit 4 and Exit 3	7120	7434–7731	3.68	Yes
Between Exit 3 and exit 2	5886	5979–6184	1.21	Yes
Between Exit 2 and Exit 1	5712	5141–5497	2.87	Yes
Southbound right onto Fayette Street	1429	1284–1428	0	Yes
Southbound through onto President Street	2673	2097–2336	6.73	No*
Southbound left onto Fayette Street	1610	1392–1592	0.45	Yes

Note: * A GEH between 5 and 10 is not considered to indicate that the model is a poor fit, but does indicate that further investigation is required.

Traffic and incident conditions were simulated along a straight and level section of a three-lane unidirectional freeway for at least 150 minutes (the software's maximum allowed simulation time). Different scenarios of incident durations and traffic intensity levels (Rho values) were generated. The Rho is a measure of traffic intensity along a segment of the freeway and is defined by the volume-to-capacity ratio (V/C). Freeway capacity was determined as 2400 vehicles per hour per lane (vphpl). Freeway capacity was defined as the traffic volume at which the throughput did not change or declines even as the input flow continually increased. For basic freeway segments with no incidents, Highway Capacity Manual (TRB, 2000) recommends an ideal capacity of 2250 to 2400 vphpl depending on free flow speed.

Traffic intensity was categorized into three levels, light ($0.25 \leq Rho \leq 0.50$), moderate ($0.50 < Rho \leq 0.80$) and near capacity ($0.80 < Rho < 1.00$). Then, incidents of various durations were generated for these traffic demand levels. Incident durations were defined as short (inc≤20 min), moderate (20 min<inc≤40 min) and long (40 min<inc≤60 min). The timed incidents were preceded by a 30-minute traffic build-up along the simulated freeway segment. The experiments covered various lane-blockage scenarios for the three-lane freeway segment. Since VISSIM cannot simulate road incidents

directly, the authors created pseudo-incidents using traffic signals. In addition, only one isolated incident per time and space were considered in the simulation. In other words, the impacts of multiple incidents on congestion and recovery time were not included in the scenarios.

A total of 121 scenarios of Rho and incident duration were generated, each with six random seeds, therefore, a total of 726 (121×6) experiments were simulated. Of the 121 scenarios, 97 scenarios involved 3 blocked lanes, 12 scenarios involved 2 blocked lanes, and 12 scenarios involved 1 blocked lane. From these 726 experiments, output values for flow, density, speed, and TRT values were derived. Various lane-blockages indicate incident severity, with more lanes closed, the incident would be more severe.

The effective Rho value for each scenario was derived as a ratio of total traffic demand for the simulation period to effective capacity for the specified incident duration. Total traffic demand was calculated for the simulation period at the specified traffic intensity or initial Rho value. Effective capacity was defined as the potential throughput for the simulation period less than the unmet demand for the incident duration at the specified traffic intensity level or original Rho value.

Determination of simulation values for TRT

Determination of the post-incident TRT was based on the pre-incident traffic conditions of density and speed.

The authors determined TRT when post-incident density values were the same as pre-incident values. Although speed values were also considered, density values were a more accurate indicator of traffic congestion along freeways, as freeways can be heavily congested even at free flow speeds. A *t*-test was then conducted to investigate if there were any significant differences between the pre- and post-incident density values. The *t*-test results presented in Table 2 verify that there is no significant difference, which means that the TRT is determined correctly in all scenarios.

Table 2. Summary results of t-test for paired two samples (density means).

Traffic intensity (Rho)	Incident duration (min)	Pre-incident density means	Post-incident density means	Sample size N	t-stat	P-value	$t$critical
0.9	10	116	117	6	-0.227	0.829	2.571
0.8	15	103	103	6	0.095	0.928	2.571
0.8	30	106	102	4	1.367	0.265	3.182
0.7	50	91	88	3	1.389	0.299	2.571

Queue dissipation time calculations using shockwave theory

The authors calculated the queue dissipation time for each three-lane-closed scenario using the following formulas

$$u_1 = \frac{q_2 - q_1}{k_2 - k_1} = \frac{0 - q_1}{k_j - k_1} \tag{1}$$

$$u_2 = \frac{q_3 - q_2}{k_3 - k_2} = \frac{q_{max} - 0}{k_c - k_j} \tag{2}$$

$$u_3 = \frac{q_4 - q_1}{k_4 - k_1} = \frac{q_{max} - q_1}{k_c - k_1} \tag{3}$$

$$Q = t_1 u_1 \tag{4}$$

$$t_2 = \frac{Q}{u_2 - u_1} \tag{5}$$

where u_1 is queue build up rate (mile/h), u_2 is queue dissipation rate (mile/h), u_3 is normal stabilization flow rate (mile/h), q_1 is pre-incident flow rate (veh/h), q_2 is incident flow rate (veh/h), q_3 is capacity flow rate (veh/h), q_4 is stabilization flow rate (veh/h),q_{max} is maximum flow capacity (veh/h), k_1 is pre-incident density (veh/mile), k_j is jam (incident) density (veh/mile), k_c is capacity (dissipation) density (veh/mile), Q is queue length (mile), t_1 is incident duration (h), t_2 is queue dissipation time (h).

When comparing simulation and shockwave, the authors considered only the case that all lanes of the freeway were closed, q_2=0. This was done in order to evaluate the worst-case scenarios for TRT of all lanes closed.

Further, the authors used simulation values for k_j, k_c, and q_{max} instead of actual field data because of two reasons. First, this would be feasible to compare the results of the shockwave theory calculations with the simulation results. Second, it would be very time-consuming and expensive to collect all scenarios of incident duration and traffic intensity. Since only non-recurring incidents were considered, it would be even more difficult to collect the actual field data.

Previous literature suggests shockwave theory calculates only queue dissipation time, which is different from TRT. In order to make a fair comparison of the simulation results and shockwave theory calculations, the authors proposed a formula to calculate the full TRT for shockwave with both the summation of queue dissipation time and the stabilization time as follows

$$t_4 = t_2 + t_3 \tag{6}$$

$$t_3 = \frac{Q}{u_3} \tag{7}$$

where t_3 is shockwave stabilization time, t_4 is shockwave recovery time.

As stated earlier, full TRT is defined as the time after incident clearance when pre-incident traffic flow conditions resume. Therefore, the full TRT is the summation of queue dissipation time and traffic stabilization time. The authors defined the queue stabilization time as the queue length divided by the stabilization rate.

RESULTS AND DISCUSSION

This section presents the outputs of the simulation scenarios and a statistical analysis of TRT for different incident and traffic demand regimes. Results of the shockwave calculations for TRT and a comparison of methods (simulation vs. shockwave) are also presented.

TRT from simulation

Presented in Table 3 are the aggregated results for 3 lanes blocked, 97 cases, each with 6 random seeds (582 experiments) of derived TRT for different scenarios of Rho and incident durations. Results for 13 of those scenarios were omitted from further analysis because the normal pre-incident condition was not attained within the 150-minute simulation period.

Table 3. Estimated TRT from traffic simulation.

Lane closure	Incident duration (min)		Short (SI) (inc≤20)				Moderate (MI) (20< inc≤40)				Long (LI) (40< inc≤60)			
	Traffic intensity (Rho)		5	10	15	20	25	30	35	40	45	50	55	60
3-lane blocked	Near capacity (NC) (0.80<Rho<1.00)	0.95	76											
		0.90	59	84	88									
		0.85	35	58	78	90	93							
	Moderate traffic (MT) (0.50<Rho≤0.80)	0.80	28	47	60	77	89	94	97					
		0.75	21	35	51	62	73	74	81	X	X			
		0.70	21	34	49	57	62	72	77	X	X	X	X	X
		0.65	16	26	35	43	51	64	68	71	73	X	X	X
		0.60	15	21	28	36	43	51	60	67	69	X	X	X
	Light traffic (LT) (0.25≤Rho≤0.50)	0.50	15	19	24	29	35	38	42	46	50	55	58	60
		0.35	12	14	15	21	23	27	28	30	32	35	38	40
		0.25	9	11	12	15	16	17	18	20	21	23	25	26

Note: X is post-incident TRT values omitted, as normal pre-incident condition is not attained during the 150 min simulation period.

The recovery time values derived from traffic simulations ranged from a high value of 97 minutes to a low value of 9 minutes, depending on the demand or traffic intensity levels and incident duration. For example, a 60-minute recovery time is associated with scenarios for NC-SI, MT-MI, and LT-LI. Scenarios for Rho 0.90 and 5 min incident, Rho 0.60 and 35 min incident, and Rho 0.50 and 60 min incident all have a recovery time of almost 60 min.

Fig. 3 presents typical density profiles for the same incident duration as traffic demand increases. As expected, the graph confirms that within the same incident duration, TRT increases nonlinearly as traffic intensity builds. Further analysis of the results suggests that as traffic intensity approaches the capacity threshold (Rho equals 1), recovery time becomes indefinite. Consequently, congestion increases as incident duration increases at all Rho (demand) values. Similarly, density profiles for different incident duration and increasing traffic intensity levels confirm that a lower incident time does not necessarily result in a lower recovery time, as TRT is a function of both incident duration and traffic intensity. Only if traffic demand is fixed the lower TRT with the lower incident time. Fig. 4shows a 3-dimensional view of the variability level in TRT associated with varying incident durations and traffic demands.

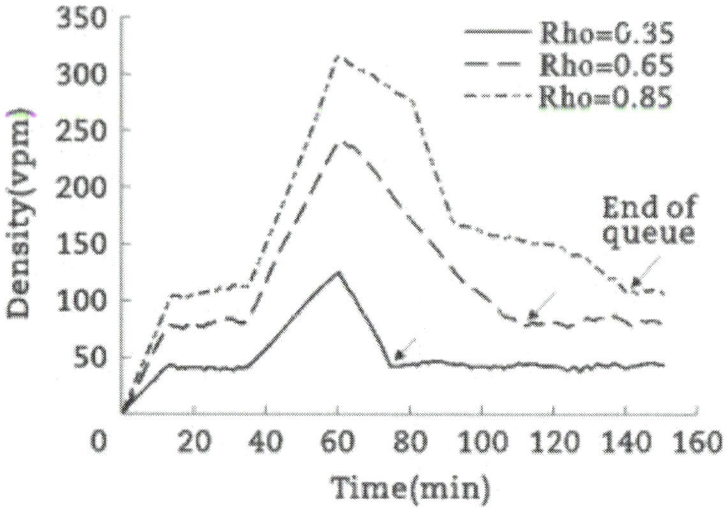

Figure 3. Simulation results of 25-min incident duration for varying demand levels.

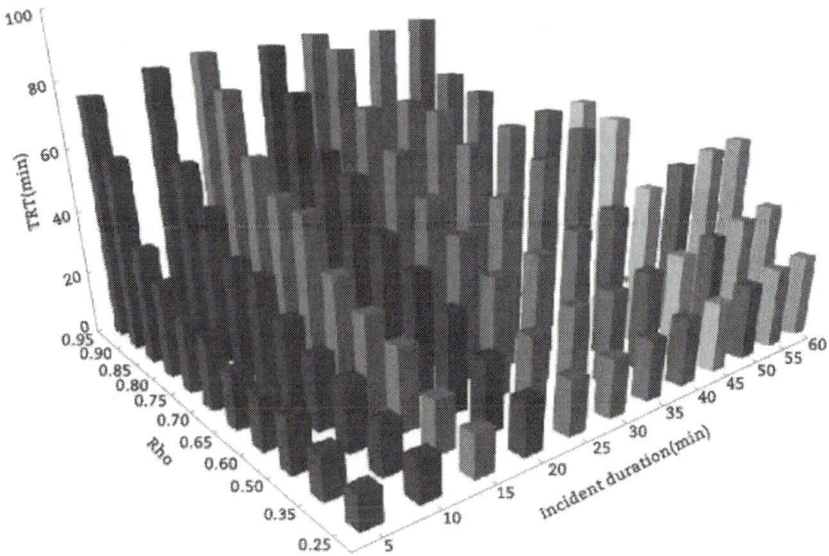

Figure 4. TRT versus Rho and incident duration.

Comparison between shockwave and simulation

TRT was also calculated by shockwave theory for the same scenarios of traffic intensity and incident duration. Table 4 presents the parameters used to calculate shockwave TRT. Table 5 presents a sample of the results and compares the estimated TRT values from simulation and from shockwave theory. In all cases, the simulated TRT values exceed the recovery time derived from shockwave methodology. On the other hand, the shockwave model's queue dissipation time is almost half of the value which was observed in the simulation.

Table 4. Parameters used in shockwave calculations.

Parameter	Value
*k*j (veh/mile)	263
*k*c (veh/mile)	132
*q*max (veh/h)	7200

Table 5. Comparison of TRT values from simulation and shockwave models (based on scenario matrix).

Scenario matrix	Incident duration t_i (min)	Original Rho	Simulation dissipation time (min)	Shockwave dissipation time (min)	Shockwave recovery time t_s (min)	Simulated recovery time t_r (min)	t_r/t_s	t_s/t_i
LT-LI	45	0.50	39	21	37	50	1.3	1
LT-LI	55	0.35	32	14	26	38	1.4	1
LT-LI	60	0.25	23	10	19	26	1.4	0
LT-LI	60	0.50	48	28	50	60	1.2	1
LT-MI	25	0.25	11	4	8	16	2.1	1
LT-MI	30	0.35	18	8	14	27	1.9	1
LT-MI	35	0.50	33	16	29	42	1.5	1
LT-SI	5	0.35	4	1	2	12	4.9	2
LT-SI	5	0.25	2	1	2	9	5.9	2
LT-SI	5	0.50	6	2	4	15	3.5	3
LT-SI	10	0.25	5	2	3	11	3.4	1
LT-SI	10	0.35	7	3	5	14	3.0	1
LT-SI	15	0.50	18	7	12	24	2.0	2
LT-SI	15	0.25	7	2	5	12	2.6	1
LT-SI	20	0.25	9	3	6	15	2.5	1
MT-LI	45	0.60	61	30	53	69	1.3	2
MT-LI	45	0.65	61	36	62	73	1.2	2
MT-MI	25	0.60	44	17	29	43	1.5	2
MT-MI	25	0.75	55	30	51	73	1.4	3
MT-MI	30	0.65	41	24	42	64	1.5	2
MT-MI	30	0.75	56	36	61	74	1.2	2
MT-MI	35	0.70	64	34	58	77	1.3	2
MT-MI	35	0.80	64	53	88	97	1.1	3

MT-SI	5	0.70	14	5	8	21	2.6	4
MT-SI	5	0.80	15	8	13	28	2.2	6
MT-SI	10	0.60	18	7	12	21	1.8	2
MT-SI	15	0.70	35	15	25	49	1.9	3
MT-SI	15	0.80	66	23	38	60	1.6	4
MT-S.	20	0.65	28	16	28	43	1.6	2
.IT-SI	20	0.75	53	24	40	62	1.5	3
NC-MI	25	0.85	79	49	80	93	1.2	4
NC-SI	5	0.85	25	10	16	35	2.2	7
NC-SI	5	0.90	37	13	22	59	2.7	12
NC-SI	5	0.95	45	19	35	76	2.2	15
NC-SI	15	0.90	81	39	66	88	1.3	6
NC-SI	20	0.85	47	39	64	90	1.4	5

The shockwave model calculations indicate that shockwave TRT approximates simulated TRT for LT-LI, MT-MI, and MT-LI scenarios. For short incidents with both off-peak (light traffic) and peak (near capacity) demand levels, TRT for almost all (except 2) cases differs by more than a 2:1 ratio for simulation values versus shockwave results.

When traffic intensity is moderate or near capacity, shockwave TRT exceeds incident duration. When traffic is light, TRT is less than incident duration. The simulation values also show varying ratios of incident to TRT depending on the variation of traffic demand. Preliminary analysis shows that the ratio of incident duration to TRT varies and when traffic demand is at low traffic (Rho<0.60) within moderate to long incident durations, TRT is approximately a factor of one. At moderate traffic with short incident duration, the factor of incident duration to TRT varies from 2 to 6. The variation in recovery time cannot be adequately described by one consistent ratio but is, in fact, determined to a certain extent based on the traffic demand at the time of the incident. As presented in Table 5, this ratio is between 1 and 15.

As stated earlier, the shockwave formula to calculate TRT was developed only for all 3 lanes closed. When only one or two lanes are closed, $q2$ is not zero and vehicles merge to other lanes. Too many assumptions need to be made to develop an accurate TRT formula. The authors decided to compare only all lanes closed in order to show the differences between results from simulation and shockwave models, which could demonstrate the superiority of simulation method.

Low traffic intensity and variations of incident duration resulted in little or no change in the TRT for partial-lane-blockage scenarios. Consequently, fewer scenarios (12 scenarios) were analyzed for the 3-lane-blocked scenarios. The recovery time incrementally increased as the number of lane closures increased from one to three lanes. Table 6compares TRT ualues for different lane-blockage scenarios.

Table 6. Sample results comparing TRT values for different lane closure scenarios.

Simulation scenario		TRT(min)		
Incident duration (min)	Original Rho	3-lane closure	2-lane closure	1-lane closure
10	0.9	86	57	28
15	0.9	93	74	42
15	0.85	78	52	26
15	0.8	60	44	17
15	0.75	51	35	16
15	0.7	49	31	11
30	0.8	85	62	22
30	0.75	71	54	19
45	0.75	72	68	33
50	0.7	65	60	26
55	0.65	64	53	30
60	0.7	*	60	32

Note: * Post-incident TRT was omitted because it was inconclusive.

Regression results

Regression analysis was employed to formulate the TRT based on traffic intensity, incident duration, and incident severity. Statistical analysis of the simulation results indicates that TRT is exponential and its natural logarithm transformation can be reasonably represented as a linear function of incident duration and traffic intensity. However, the authors found that when scenario is defined within the specified ranges of traffic intensity level, a better model is obtained using a regression model to estimate the coefficients of this relationship.

Four different regression models were explored, aggregated for all scenarios, based on traffic intensity levels as defined above, and for each lane blockage scenario separately. The developed regression model is as follow, and the results are presented in Table 7.

$$\ln(TRT)=\alpha+\beta t_1+\gamma Rho+\theta L \quad (8)$$

where L is the ratio of lanes closed, α is a constant, β , γ,θ are coefficients.

Table 7. Regression results for aggregated, traffic intensity, and lane blockage scenarios.

Regression model		α	β	γ	θ	R^2	Adj R^2	N
All scenarios aggregated		−0.406 (0.1705)	0.0216 (8.18E−13)	2.634 (1.49E−22)*	1.833 (2.01E−13)	0.662	0.653	121
Traffic intensity level	Near capacity	−0.382 (0.784)	0.024 (0.0001)	3.179 (0.051)	1.188 (0.009)	0.480	0.443	47
	Moderate traffic	−1.766 (0.016)	0.025 (7.58E−08)	3.679 (8.98E−05)	2.392 (2.81E−10)	0.676	0.655	50
	Low traffic	1.512 (5.94E−12)	0.016 (6.77E−12)	3.113 (1.01E−08)	·	0.942	0.882	24

Note: * The numbers in parenthesis are P values of coefficients. P values less than 0.05 are considered significant at the 95% level of confidence Interval.

All regression models with 0 intercept show a strong correlation among all the variables, with high adjusted R^2 values. It suggests that over 85 percent of the variance in post-incident traffic recovery can be explained by the variables of traffic intensity, incident duration, and the proportion of lane closure. Note that "L" in the regression formula is the proportion of lane closed, e.g., when one lane out of the three lanes of the freeway is closed, L is 1/3 or 0.333.

Model validation

The developed simulation-based regression model was validated using real-world data. The authors obtained incident data from the Regional Integrated Transportation Information System (RITIS) in CATT Laboratory at the University of Maryland, College Park (RITIS, 2012). RITIS integrates existing transit and transportation management data from different sources. The main RITIS functions are the real-time fusion and exchange of regional transportation data, and data archiving. The major data sources are traffic cameras. The authors extracted traffic data for incidents on I-83 and I-695 that occurred from June to September, 2011, based on the cameras located on these two urban freeways. Number of lanes closure, incident duration, and Rho were extracted from the database and TRT was derived based on the database speed information in the same way it was calculated for simulation results. TRT was also predicted using the developed regression formulas as presented in Table 8. The authors compared the predicted TRT using regression analysis with the observed ones. The authors excluded row 19 in Table 8 since it did not seem to be realistic to have a 10-minute observed TRT when the traffic was almost at capacity (Rho=0.99) and the incident duration was 41 min. It could be a data collection error. Columns in Table 8 present Rho, proportion of the lanes closure, incident duration, observed TRT, and predicted TRT.

Table 8. Comparison of observed and predicted TRT values.

Case	Rho	Lane closure proportion	Incident duration (min)	RITIS observed TRT (min)	Predicted TRT (min)
1	0.53	0.67	10	15	14.81
2	0.18	0.33	109	21	26.83
3	0.46	0.33	30	45	11.28
4	0.18	0	22	3	2.61
5	0.09	1	72	20	25.43
6	0.31	0.67	20	13	9.72
7	0.3	1	23	22	16.38
8	0.5	0.33	37	18	14.72
9	0.36	0.67	31	59	14.25
10	0.3	1	21	15	15.92
11	0.14	0.33	33	22	4.85
12	0.57	0.33	26	20	13.98
13	0.58	0.67	18	17	20.04
14	0.94	0.33	44	51	58.02
15	0.4	0.67	14	21	11
16	0.56	0.33	15	37	10.8
17	1.02	0.33	4	45	31.2
18	0.67	1	19	69	43.32
19	0.99	0.67	41	10	103.44
20	0.66	0.67	30	45	31.82
21	0.68	0.33	53	48	33.92
22	0.7	0.33	30	25	22.3
23	0.92	0.33	16	24	30.23
24	0.3	0	4	3	2.55
25	0.31	1	24	34	17.36
26	0.47	0.67	72	45	46.25
27	0.55	0.33	51	11	22.54
28	0.66	0.67	20	24	26.49
29	0.33	0	1	2	2.57
30	0.51	1	36	38	39.45
31	0.61	0.33	15	18	12.41

The authors conducted a single regression between observed and predicted TRT. The regression results indicate that the predicted TRT using the regression model developed by the authors explains 82% of the observed data (R^2=0.821) as presented in Table 9. The authors also used a Wilcoxon rank-sum test, which was a non-parametric test, to compare the observed and predicted TRT ualues. The test results showed no significant difference between the probability distributions of the predicted and the observed TRT with 95% confidence level (Z=-2.451, Sig. = 0.014).

Table 9. Regression results for observed TRT versus predicted TRT.

Regression	Predicted TRT	R^2	Adj. R^2	N	P-value
Coefficient	1.17	0.821	0.786	30	2.30E-12

Based on the Wilcoxon test and regression results, the authors' simulation-based regression model provides a reliable formula to predict TRT for different flow regimes on urban freeways. The model can be utilized for urban freeways other than I-83, within the defined clonstraints, since I-695's predicted TRT was close to the observed TRT.

CONCLUSIONS

This study proposed a methodology for estimating post-incident traffic recovery time (TRT). Traffic simulation was utilized to evaluate how different combinations of demand (V/C), incident duration, and incident severity (lane closure proportion) affected incident recovery time. Simulation was utilized to have all different combinations of traffic demand (or intensity, V/C of 0.1 to 1), incident duration (5 to 60 minutes), and incident severity (the proportion of lanes closure). A total of 726 simulations were completed. Finding all these combinations was almost impossible from real-world data.

The authors then applied the simulation output results for speed, density, and flow to well-known analytic delay prediction formulas to compare the results of TRT. In addition, the authors developed a regression model that can reasonably estimate recovery time based on 3 primary variables: traffic intensity, incident duration, and lane closure proportion. TRT is defined as the period elapsing after incident clearance when traffic returns to pre-incident conditions.

Since full TRT estimation has not been widely explored in literatures, this research is relevant and timely to the transportation industry and the transportation management center (TMC) responsible for smooth and efficient freeway and highway operations. Most traffic managers have postulated that the post-incident TRT exceeded the actual incident duration by a fixed factor.

The traffic simulation results indicate that congestion increases as incident duration increases at all demand levels but increases at faster rates for higher traffic demand. However, recovery time becomes indefinite as traffic intensity approaches capacity threshold. This suggests that TMCs should implement alternate incident management strategies once a certain demand threshold is reached. A regression model was developed to estimate TRT using the simulation results. The regression model was applied to different combinations of traffic intensity level, incident durations, and lane blockage. All regression models show a very strong positive relationship between a natural log of TRT and incident time, demand, and incident severity. The validity of the simulation-based regression model results was successfully tested using the collected field data.

Comparative analysis of the simulation- and shockwave-derived TRT values suggests that the simulation model offers some advantages over the traditional shockwave model. Except for the low traffic (off-peak) scenarios, the simulation-based TRT estimates consistently exceeded the shockwave-derived estimates. In those low traffic scenarios, the shockwave-derived TRT approximated the simulation-derived results. The shockwave-derived results were consistently lower for both queue dissipation and recovery time.

Engineers and safety officials can apply the developed formulas within the defined constraints to estimate the TRT after an incident along an urban freeway instead of using a fixed factor, such as 4. The simulation model is a better alternative because it can model various incident durations and scenarios.

The ability to utilize extensive simulation data in scenario analysis will enhance the management agencies' ability to quantify the impact of congestion and delay on the highway network. The regression formula for determining post-incident TRT will enable traffic management personnel to systematically ascertain the magnitude of traffic congestion conditions along highways. In addition, it will be possible to reasonably estimate the effect of proportional lane closures and increase traffic intensity on congestion buildup.

The authors considered only three factors, traffic intensity, incident severity, and incident duration that affect TRT. While other factors such as road geometry could affect TRT, the authors tried to develop a very simplistic method to be usable by practitioners. An extension to this model could add more factors and also account for the rubbernecking effect. Another extension to this study would consider the impact of multiple incidents on congestion and recovery time.

ACKNOWLEDGMENTS

The authors would like to acknowledge and thank the following agencies for their funding support and input throughout the study, the Maryland State Highway Administration Office of Traffic and Safety, National Transportation Center at Morgan State University, and the Federal Highway Administration through the Dwight Eisenhower Fellowship Program.

REFERENCES

1. Boyles, S., Waller, S.T., 2007. A stochastic delay prediction model for real-time incident management. Institute of Transportation Engineering 77 (11), 18-25.
2. Chang, E.C., Gang, L., Steven, R., 2006. Performance evaluation of CHART - the real time incident management system in Year 2005. Maryland State Highway Administration, Baltimore.
3. Chien S.I.J., Chowdhury, S.M., 2000. Simulation-based estimates of delays at freeway work zones. In: Proceedings of the Transportation Research Board 80th Annual Meeting, Washington DC, 2000.
4. Dowling, R., Skabardonis, A., Halkias, J., et al., 2004.
5. Guidelines for calibration of microsimulation models: framework and applications. Transportation Research Record (1876), 1-9.
6. Garber, N.J., Hoel, L., 2002. Traffic & Highway Engineering 3rd edition. Thomson Learning, Stamford.
7. Garib, A., Radwan, A.E., Al-Deek, H., 1997. Estimating magnitude and duration of incident delays. Journal of Transportation Engineering 123 (6), 459-466.

8. Giuliano, G., 1989. Incident characteristics, frequency, and duration on a high volume urban freeway. Transportation Research Part A: General 23 (5), 387-396.
9. Hadi, M.A., Sinha, P.K., Wang, A., 2007. Modeling reductions in freeway capacity due to incidents in microscopic simulation models. Transportation Research Record (1999), 62-68.
10. Jha, M., Gopalan, G., Garms, A., et al., 2004. Development and calibration of a large-scale microscopic traffic simulation model. Transportation Research Record (1876), 121-131.
11. Knoop, V.S., 2010. Stochastic incident duration: impact on delay. In: Proceedings of Transportation Research Board 89th Annual Meeting, Washington DC, 2010.
12. Li, J., Lan, C.J., Gu, X., 2006. Estimation of incident delay and its uncertainity on freeway networks. Transportation Research Record (1959), 37-45.
13. Lindley, J.A., 1987. A methodology for quantifying urban freeway congestion. Transportation Research Record (1132), 1-7.
14. Morales, J.M., 1987. Analytical procedures for estimating freeway traffic congestion. Transportation Research Circular (344), 45-49.
15. Oketch, T., Carrick, M., 2005. Calibration and validation of a micro-simulation model in network analysis. In: Proceedings of the Transportation Research Board 85th Annual Meeting, Washington DC, 2005.
16. Olmstead, T., 1999. Pitfall to avoid when estimating incident-induced delay by using deterministic 453 queuing models. Trasnportation Research Record (1683), 38-46.
17. PTV, 2000. VISSIM 4.3 User Manual. Planning Transport Verkehr, Karlsruhe.
18. ITIS, 2012. Available at http://www.cattlab.umd.edu/?portfolio=ritis (Date accessed: 31 January 2012).
19. Saka, A.A., Jeihani, M., James, P.A., 2008. Estimation of post-incident traffic recovery time under different flow regimes. Research Report MD-09SP708B4L, Maryland State Highway Administration, Baltimore.
20. Skabardonis, A., Geroliminis, N., 2004. Development and application of methodologies to estimate incident impacts. In: Proceedings of the 2nd International Congress on Transportation Research, Berkeley, 2004.
21. Sullivan, E.C., 1997. New model for predicting freeway incidents and incident delays. Journal of Transportation Engineering 123 (4), 267-275.
22. Toledo, T., Koutsopoulos, H.N., 2004. Statistical validation of traffic simulation models. Transportation Research Record (1876), 142-150.
23. TRB, 2000. Highway Capacity Manual 4th Edition. Transportation Research Board, Washington DC.

24. Wiedemann, R., 1974. Simulation des Straßenverkehrsflusses. Schriftenreihe des, Institut für erkehrswesen. Universität Karlsruhe, Karlsruhe.
25. Wiedemann, R., 1991. Modelling of RTI-elements on multi-lane roads. In: Drive Conference, Brussels, 1991.

CITATION

Mansoureh Jeihani, Petronella James, Anthony A. Saka, Anam Ardeshiri, Traffic recovery time estimation under different flow regimes in traffic simulation, Journal of Traffic and Transportation Engineering (English Edition), Available online 7 August 2015, ISSN 2095-7564, http://dx.doi.org/10.1016/j.jtte.2015.08.001.

CHAPTER 4

Development of an Open-Source Road Traffic Noise Model for Exposure Assessment

John Gulliver[1], David Morley[1], Danielle Vienneau[2, 3], Federico Fabbri[1], Margaret Bell3, Paul Goodman[4], Sean Beevers[5], David Dajnak[5], Frank J Kelly[5], Daniela Fecht[1]

[1] The UK Small Area Health Statistics Unit (SAHSU), MRC-PHE Centre for Environment & Health, Imperial College London, W2 1PG, London, UK
[2] Department of Epidemiology and Public Health, Swiss Tropical and Public Health Institute, 4002 Basel, Switzerland
[3] University of Basel, 4003 Basel, Switzerland
[4] School of Engineering and Geosciences, Newcastle University, NE1 7RU, UK
[5] Environmental Research Group, King's College London, SE1 9NH, UK

HIGHLIGHTS

- Adaptation of the Calculation of Road Traffic Noise method for exposure assessment.
- Freely available open-source software (R with PostgreSQL and GRASS GIS).
- Model estimates compared well to noise measurements (r: ~0.85–0.95).
- Noise level exposures modelled for 8.61 million London residents (2003–2010).
- Over 1 million residents exposed to high daytime and night-time noise levels.

ABSTRACT

This paper describes the development of a model for assessing **Tra**ffic **N**oise **Ex**posure (TRANEX) in an open-source geographic information system. Instead of using proprietary software we developed our own model for two main reasons: 1) so that the treatment of source geometry, traffic information (flows/speeds/spatially varying diurnal traffic profiles) and receptors matched as closely as possible to that of the air pollution modelling being undertaken in the TRAFFIC project, and 2) to optimize model performance for practical reasons of needing to implement a noise model with detailed source geometry, over a large geographical area, to produce noise estimates at up to several million address locations, with limited computing resources. To evaluate TRANEX, noise estimates were compared with noise measurements made in the British cities of Leicester and Norwich. High correlation was seen between modelled and measured $L_{Aeq,1hr}$ (Norwich: $r = 0.85$, $p = .000$; Leicester: $r = 0.95$, $p = .000$) with average model errors of 3.1 dB. TRANEX was used to estimate noise exposures ($L_{Aeq,1hr}$, $L_{Aeq,16hr}$, L_{night}) for the resident population of London (2003–2010). Results suggest that 1.03 million (12%) people are exposed to daytime road traffic noise levels ≥ 65 dB(A) and 1.63 million (19%) people are exposed to night-time road traffic noise levels ≥ 55 dB(A). Differences in noise levels between 2010 and 2003 were on average relatively small: 0.25 dB (standard deviation: 0.89) and 0.26 dB (standard deviation: 0.87) for $L_{Aeq,16hr}$ and L_{night}.

INTRODUCTION

There is growing concern on the effects of noise pollution on health (WHO, 2009). Environmental noise exposure is associated with annoyance (Babisch et al., 2013 and Clark et al., 2012), sleep disturbance (WHO, 2011), cognitive ability in schoolchildren (Evans and Hygge, 2007 and Clark et al., 2013), and health impacts, especially cardiovascular conditions and risk factors (Babisch and van Kamp, 2009,Hansell et al., 2013, Sorensen et al., 2012 and Stansfeld and Matheson, 2003). Exposure to environmental noise is ubiquitous and increasing in terms of road traffic noise and the reduction of the night-time quiet period (Hammer et al., 2014). Traffic-related noise is said to account for over 1 million healthy years of life lost annually to ill health and may lead to a disease burden that is second only in magnitude to that from air pollution (WHO-JRC, 2011).

Estimates of environmental noise exposures were required for a series of epidemiological analyses forming part of the 'Traffic and Health in London (TRAFFIC)' study (2011–2014) (http://www.kcl.ac.uk/lsm/

research/divisions/aes/research/ERG/research-projects/traffic/index.aspx), funded by the Natural Environment Research Council (NERC) Environmental Exposure and Health Initiative (EEHI). The epidemiological analyses were designed to look at relationships of a range of air pollution metrics (e.g. particulate matter, nitrogen oxides, nitrogen dioxide, ozone) and noise levels with health outcomes in children and adults, including cardiovascular and respiratory mortality and hospital admissions, adverse birth outcomes (low birth weight and pre-term delivery), primary care data on diseases and consultations, and risk factors and vascular markers of diseases in children, over different periods between 2003 and 2010. In addition to confounders (e.g. smoking, deprivation) the epidemiological analyses of air pollution may control for noise levels and vice versa.

This paper describes the development of a **TRA**ffic **N**oise **EX**posure model (TRANEX) and its implementation for exposure assessment in the TRAFFIC study. Instead of using proprietary software (e.g. CadnaA, SoundPLAN) we developed our own model for two main reasons: 1) so that the treatment of source geometry, traffic information (flows/speeds/spatially varying diurnal traffic profiles) and receptors matched as closely as possible to that of the air pollution modelling being undertaken in the TRAFFIC project, and 2) to optimize model performance for practical reasons of needing to implement a noise model with detailed source geometry, over a large geographical area, to produce noise estimates at up to several million address locations, with limited computing resources. We present an evaluation of the model using comparisons of estimated noise levels with noise level measurements from the UK. We also describe the modelled noise exposures at postcode locations and for the resident population of London for an example year (2008) and changes in modelled noise exposures over the study period (2003–2010).

MATERIALS AND METHODS

As the model was developed primarily for application to London we adopted the Calculation of Road Traffic Noise method (CoRTN) (Department of Transport, 1988). The CoRTN method is used in the UK for strategic noise mapping (DEFRA, 2008 and HMSO, 2006) and has been implemented as an optional noise calculation method in leading proprietary software such as SoundPLAN (http://www.soundplan-uk.com) and CadnaA (http://www.datakustik.com/en/products/cadnaa). The CoRTN method is also used as the UK primary noise calculation methodology for new road schemes (http://www.dft.gov.uk/ha/ standards/ dmrb/vol11/section3/hd21311.pdf). The study area (Fig. 1) is the area inside the M25 motorway that includes Greater London and the surrounding area, henceforth referred to as London.

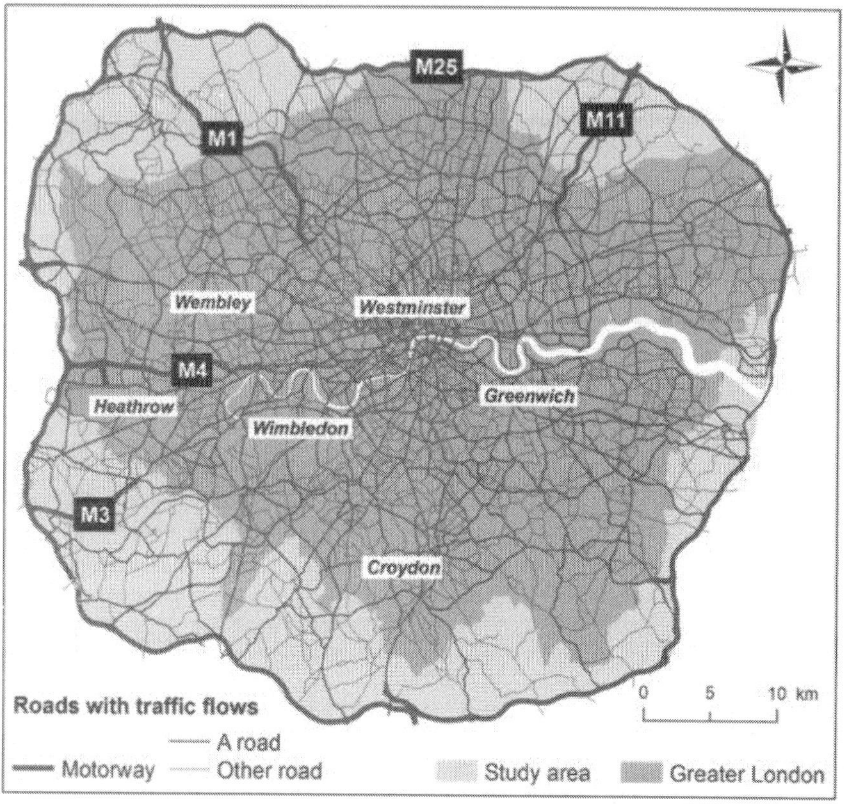

Figure 1. Study area showing the road network used in TRANEX (source data: London Atmospheric Emissions Inventory (LAEI), 2010).

The CoRTN method

Traffic noise estimates are calculated in CoRTN as one-hour, A-weighted L_{10} (dB) (i.e. the noise level that is exceeded 10% of the time), denoted as $L_{A10,1hr}$, and $L_{10,18hr}$ for the period 06:00 until 0:00.

The CoRTN method includes the following terms:

$$L_{A10,1hr}=L_0+\Delta_f+\Delta_g+\Delta_p+\Delta_d+\Delta_s+\Delta_c+\Delta_a+\Delta_r \quad (1)$$

where $L0$ is the basic noise level calculated at 3.5 m from the kerbside, at 0.5 m above ground level; Δf is the correction for traffic speed and the percentage of heavy vehicles; Δg is the adjustment for the gradient of a road section; Δp is the road surface correction; Δd is the slant distance between the road (source) and receptor; Δs is the correction for shielding (i.e. barriers) between a road (source) and receptor; Δc is ground cover attenuation; Δa is the correction for the angle of view of the road; Δr is the correction for reflections from buildings on the opposite side of the façade.

$$L_0 = 42.2 + 10\, Log_{10}\, q\ dB(A) \tag{2}$$

where q is the hourly traffic flow;

$$\Delta_f = 33Log_{10}(v + 40 + 500/v) + 10Log_{10}(1 + 5p/v) - 68.8\, dB(A) \tag{3}$$

v is the average hourly traffic speed and p is the percentage of heavy vehicles on each road section;

$$p = 100f/q \tag{4}$$

f is the hourly flow of heavy vehicles;

$$\Delta_g = 0.3\, G \tag{5}$$

where G is the gradient of the road – only applicable to uphill flow on single carriageways;

Δp: a value of -1 dB and is applicable to bituminous impervious road surfaces for roads with vehicle speed less than 75 km/h; pervious road surfaces have a value of -3.5 dB;

$$\Delta_d = -10Log_{10}(d'/13.5)\ dB(A) \tag{6}$$

d' is the shortest slant distance between a source and receptor;

$$d' = \left[(d + 3.5)^2 + h^2\right]^{0.5} \tag{7}$$

where d is the shortest distance between source and receptor; h is the difference in height above ground between the source and receptor (i.e. effective source position); the distance correction (Δd) is only applicable where $d \geq 4$m (i.e. beyond the distance from the kerbside where the basic noise level is calculated); d is given a value of 4 m where it is measured to be < 4 m;

Δs is the correction for barriers between the source and receptor; effectively this is a lengthening of the shortest slant distance; where the path between the source and receptor is obstructed, the distance between the source and receptor is, for example, the sum of 1) the distance from the source to the top of a building (i.e. the diffracting edge) and 2) the distance from the top of a building to the receptor [further details are available from: http://resource.npl.co.uk/acoustics/techguides/crtn/];

Δc is the correction for ground cover correction:

$$\text{For } 0.75 \leq H < (d+5)/6;\ \Delta_c = 5.2\, I\, Log_{10}[(6H - 1.5)/(d+3.5)] \tag{8}$$

$$\text{For } H < 0.75;\ \Delta_c = 5.2\, I\, Log_{10}[3/(d+3.5)] \tag{9}$$

$$\text{For } H \geq (d+5)/6;\ \Delta_c = 0 \tag{10}$$

where H is the average height of propagation (m), d is the distance from the edge of the roadside to the receptor, and I is based on the different percentage s of absorbent ground between the source and the receptor;

$$\Delta_a = 10 Log_{10}(\theta/180)\ \mathrm{d}B(A) \tag{11}$$

θ is the angle of the view of the road in degrees;

$$\Delta_r = 1.5(\theta_s/\theta)180^\circ\ \mathrm{d}B(A) \tag{12}$$

θ_s is the sum of angles by all the reflecting facades (i.e. individual buildings) on the opposite side of the road to the receptor.

The CoRTN method also includes further corrections for roads with low hourly flow ($50 \leq q < 200$), and corrections for mixed ground cover.

Finally, the noise level from all sources ($L1, L2, \ldots Ln$) can be combined using the following procedure:

$$L = 10\, Log_{10}\left[\sum_{1}^{n} 10^{Ln/10}\right] \mathrm{d}B(A) \tag{13}$$

where L is the predicted noise level from n noise sources.

Further information can be found in the CoRTN manual (Department of Transport, 1988) and other implementations of the method (e.g. Pamanikabud and Tansatcha, 2003).

Data

For implementation of the noise model in London we used the most detailed data sets available for traffic information (i.e. composition, speed, diurnal varying traffic profiles for different parts of London), land cover, road geography, building heights, and receptors (i.e. postcodes and addresses). Table 1 provides a list of data types, data sources, and their spatial resolution (i.e. accuracy).

Table 1. Description of data used in TRANEX.

Variable	Data type	Provider	Source	Spatial resolution
Traffic flow	Point	King's College London	London Atmospheric Emissions Inventory (LAEI)	10 m traffic source points
Topographic Layer (including buildings and land cover)	Polygon	Digimap® (www.digimap.edina.ac.uk)	Ordnance Survey (OS) MasterMap™	<1 m
Integrated Transport Network (ITN)	Line	Digimap® (www.digimap.edina.ac.uk)	Ordnance Survey (OS) MasterMap™	<1 m
Building Heights	Polygon	Landmap (www.landmap.ac.uk)	LiDAR survey, high-resolution aerial photography	±0.5 m with 95% confidence limits
Land-Form PANORAMA Digital Terrain Model (DTM)	Regular grid	Digimap® (www.digimap.edina.ac.uk)	Ordnance Survey (OS) contours	Scale:1:50000; Horizontal resolution: 50 m Vertical resolution: 1 m
Postcodes	Point	King's College London	Ordnance Survey (OS)	<1 m

Pre-processing of spatial data was undertaken in a geographic information system (GIS) (ArcGIS v 10.0, ESRI, Inc., Redlands, California). Information on buildings and land cover (Topographic Layer) and the Integrated Transport Network (ITN) form part of Ordnance Survey's 2009 version of MasterMap™ (MM) (see Table 1). We downloaded these data from Digimap® under the agreement for use in teaching and research in higher education. Although more recent updates are available we assumed this version to best reflect our study period.

The ITN provides detailed road network information including road type and information on one-way streets. Traffic source data (i.e. 10 m points along roads) are from the London Atmospheric Emissions Inventory (LAEI) (LAEI, 2010). Information on one-way streets and road tunnels in the ITN were linked to traffic source points. We used information from the ITN to exclude road tunnels from TRANEX.

Road traffic flows in the LAEI are represented using annual averaged daily traffic (AADT) data. The AADT data for each link has been calculated in accordance with that used in the LAEI 2008, and is described in Beevers et al. (2009). There are a total of 2028 manual classified count (MCC) sites within the LAEI area used to allocate traffic to the network – a total of ~63000 road links. Not all sites are counted each year, and of the 2028, 690 were counted in 2010. The MCC data are based upon only one day of observations, so to minimise the effect of specific local events introducing outliers into the dataset, the 2010 data has been added to a series of MCC data (extending back to 1999) and smoothed using a LOcal regrESSion (LOESS) smoothing function. Furthermore, as the MCC data only cover a 12-h weekday period (07:00 to 19:00), these data were then expanded to provide counts for each hour of each day in 2010 (including weekends and overnight hours), using hourly average automatic traffic count and automatic number plate recognition data. The resulting annual hourly dataset was then averaged by link to provide AADT estimates for cars, motorcycles, buses, light goods vehicles, taxis, and 6 types of rigid and articulated heavy goods vehicles. MCC sites with data from all years were used to calibrate the 2010 AADT to produce flows and speed estimates for other years (2003–2009).

Vehicle speeds for the major road network in the LAEI 2010 are based upon a combination of TrafficMaster GPS derived and Moving Car Observer (MCO) speeds. MCO speed is observed using a vehicle travelling at the average speed of the traffic, whereas the 'TrafficMaster' speed has been derived from a GPS-based vehicle tracking system using 2009/2010 observations, and averaged into overnight (22:00–06:00), AM (07:00–09:00), inter (10:00–15:00), PM (16:00–18:00) and evening (19:00–21:00) periods of the day. GPS speed was available for approximately 62% of the LAEI 2010 major road links with MCO data covering the remaining roads.

In order to obtain hourly traffic flows for 10 m traffic source points we combined information on traffic composition from the LAEI into two vehicle categories: light duty vehicles (LDV) and heavy vehicles (HV). From this we calculated hourly information on percentage of HV for the calculation of p in the CoRTN method (equation (4)). Average vehicle speed for each traffic source point also came from the LAEI.

Building height data (i.e. LiDAR) for all built-up areas within the M25 were downloaded from the Landmap website (http://www.landmap.ac.uk; accessed on 6th August, 2013). For some outlying areas close to the M25, building height data is not available because this data only covers urban areas. MM buildings were considered to better represent the buildings within our study period and to be of better spatial resolution than buildings from Landmap. We therefore assigned building heights to the nearest MM building within a 20 m radius of each building height location. MM buildings that could not be assigned a building height (e.g., no MM building within 20 m of building height data, missing building heights in Landmap) were assigned a default building height of 10 m if the footprint of a MM building was ≥15 m^2. Small buildings such as bus shelters, porches, garages etc. potentially cause problems in the definition of building facades and noise calculations (see Fig. 2). All buildings <15 m^2 were therefore deleted (see Fig. 2). Building heights were converted into a 0.5 m × 0.5 m grid of buildings attributed with heights for viewshed analysis (i.e. for the reflections calculation).

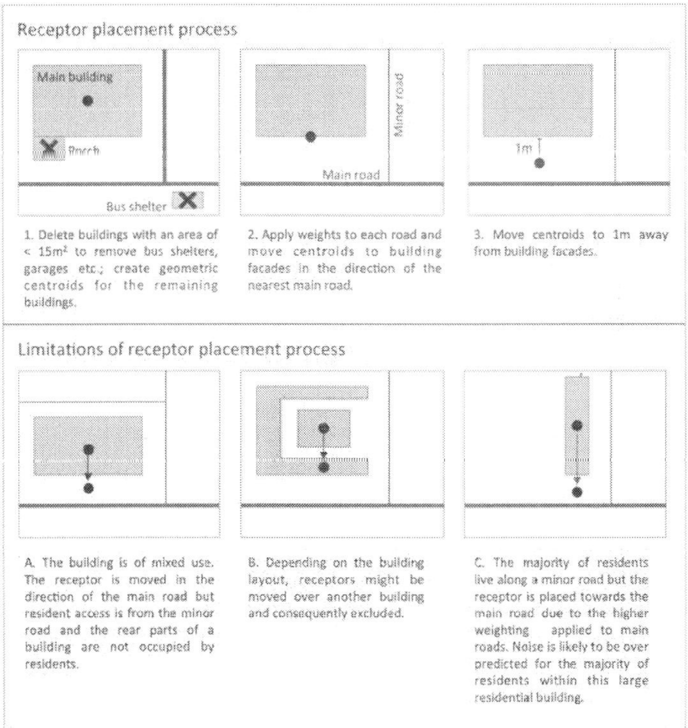

Figure 2. Procedures used to generate receptors from building polygons and some of their limitations.

For generation of receptors (i.e. address or postcode locations) a geometric centroid was created for each MM building. Each receptor was then moved to 1 m from the facade on the side of the building closest to the nearest road section with traffic information. Fig. 2shows how this was achieved and also shows situations where this automated method of moving receptors to facades does not work.

Postcodes and address points were intersected with buildings and subsequently linked to receptors using a unique building identifier. Typically there are ~15 addresses associated with each postcode. Point locations for postcodes are the geometric centroids of the address locations associated with each postcode. Each postcode is attributed with a headcount using data from the 2011 census. There are 189531 postcodes, ~3 million address locations, and a population of 8613526 in the study area. In this study we only present results related to postcode locations. Address locations are used for forthcoming individual-level health analyses.

Modifications to the CoRTN method

TRANEX broadly follows the CoRTN method with some modifications for the treatment of source geometry, the calculation of path distance, traffic on minor roads, road surfaces, tunnels, and gradients along roads. We also add in standard noise metrics ($L_{Aeq,1hr}$, $L_{Aeq,16hr}$, L_{night}) as specified in the European Noise Directive (END; European Directive 2002/49/EC).

$L_{Aeq,1hr}$ is calculated from the following empirical relationship described in Abbott and Nelson (2002):

$$L_{Aeq,1hr} = 0.94\, L_{10,1hr} + 0.77\ \mathrm{d}b(A) \tag{14}$$

TRANEX then produces $L_{Aeq,16hr}$ and L_{night} by averaging $L_{Aeq,1hr}$ from the hours 07:00 – 22:00 and 23:00 – 06:00, respectively.

For minor roads we used a fixed value of 600 vehicles day^{-1} based on the magnitude of manual counts undertaken during noise measurements, and MCC data made available by Norwich City Council and available in the LAEI. Counts were proportionally assigned to minor roads for each hour of the day using the diurnal traffic profile associated with the nearest main road in the LAEI.

The CoRTN method calculates the shortest path distance between each source and receptor along the line that bisects the angle subtended by each road section. In other words, the shortest path is taken as a line from the mid-point of each road section to each receptor. In TRANEX we use traffic information assigned to traffic source points. The shortest path is created from each of these points to receptors.

Fig. 3 shows the pre-processing steps to select traffic source points related to each receptor and the subsequent creation of ray-paths as the basis for calculating the propagation terms.

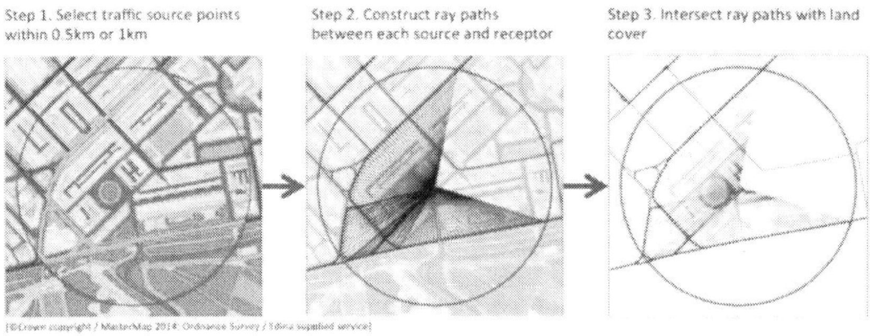

Figure 3. Construction of ray paths for the distance, shielding and ground cover corrections.

A circular buffer of 500 m is created around each receptor and intersected with the traffic source points (Step 1; Fig. 3). If no traffic source points fall within the 500 m buffer, a second buffer with a radius of 1000 m is created and the operation is repeated. This is to ensure that the nearest main road is included in the noise calculation, which is especially important over open, flat terrain as noise from heavily trafficked roads at distances of ~1 km can sometimes be heard depending on meteorology. A ray-path using straight-line geometry is created between the receptor and each traffic source point (Step 2;Fig. 3). Each ray-path is subsequently intersected with buildings and land cover for the corrections due to screening and ground cover (Step 3; Fig. 3).

In this study all receptors were given a height of 4 m above ground. For each receptor, the noise level is the combination (Equation (13)) of noise levels predicted for each pair of source-receptor points with the basic noise level adjusted for the propagation terms listed in Section 2.1. We were unable to characterise road surfaces in London so we treated all roads as bituminous impervious. We also were unable to determine the direction of flow on single carriageways so we did not include the CoRTN term for the gradient correction. Traffic in tunnels is excluded from TRANEX.

TRANEX was developed for exposure assessment (i.e. postcode/address locations), thus it is not directly usable for noise mapping (i.e. predicting over a regular grid of receptors). Fig. 4 shows the workflow of the CoRTN method applied in TRANEX including L_{Aeq} metrics and a fixed 'low flow' calculation for minor roads.

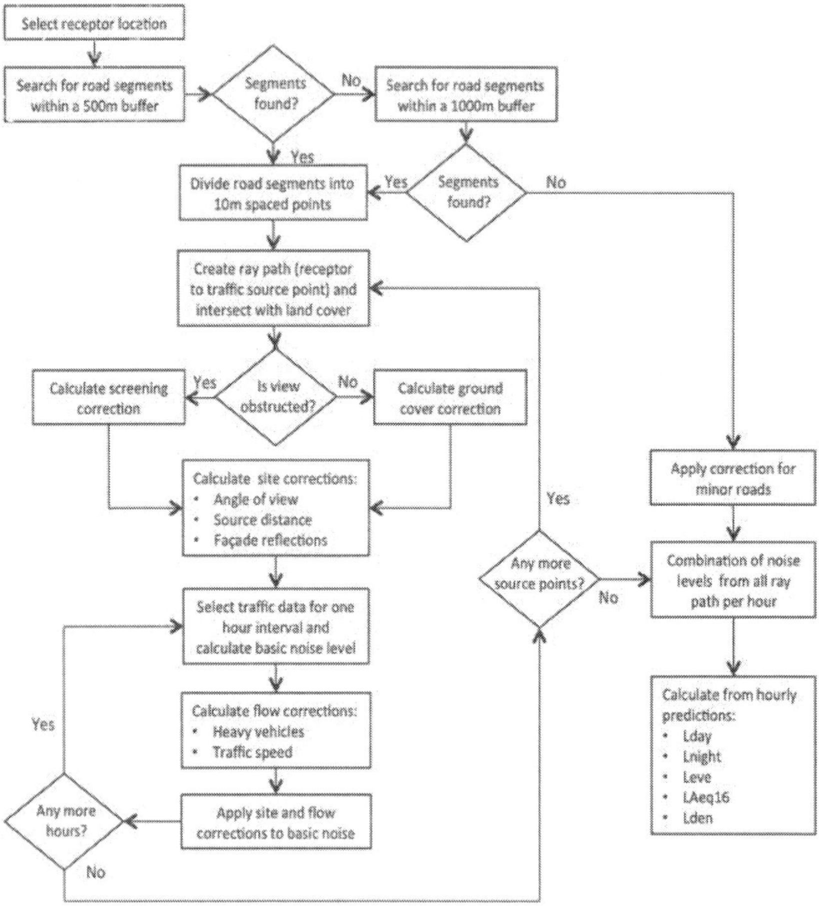

Figure 4. Workflow applied in TRANEX.

Development of software

The modified CoRTN method has been implemented using open-source software. Calculations are carried out in a PostgreSQL[1] database with GIS functions provided by the PostGIS[2] extension. In terms of processing speed for spatial operations, PostGIS has greater performance than other common desktop GIS applications due to its use of efficient spatial indexing. This is of particular importance for this application of TRANEX which uses detailed land cover polygons on a city-wide scale precise to ~1 m as an input. As such, PostGIS provides an effective environment in which to handle large vector data sets. The CoRTN method also includes a correction based on the proximity of sound reflective surfaces (building façades). This is achieved via viewshed analysis of buildings within a 50 m radius of a receptor point. As viewsheds cannot be calculated by PostGIS, GRASS GIS[3] is used and the derived representation of visible building facades was imported into the PostgreSQL database. TRANEX is controlled by a script in the R software[4] using the PostgreSQL[5] and spgrass6[6] packages to call the database and viewshed functions, respectively.

TRANEX is implemented using two R scripts: 1) a pre-processor to generate receptors 1 m from the façade of each residential building using a set of rules for best placement of receptors, and 2) a script that contains the noise model which calls PostGIS and GRASS functions. On running the R script, TRANEX proceeds on a receptor point-by-point basis. The estimated noise metrics are stored both as an ArcGIS shapefile and within the PostgreSQL database.

Model evaluation

While air pollution is routinely monitored in national networks, there is no routine monitoring for noise and data are generally not publicly available. To evaluate the performance of TRANEX, we utilised data on noise measurements collected as part of previous studies undertaken by the authors in the EU funded 5th Framework Program HEAVEN (Healthier Environment through the Abatement of Vehicle Emissions and Noise, in 2002) and HEARTS (Health Effects and Risks to Transport Systems, in 2005) projects in Leicester, UK, and data collected in 2014 to coincide with on-going air pollution monitoring being undertaken by the authors in Norwich, UK. Fig. 5 shows maps of the areas covering noise measurements used in this study.

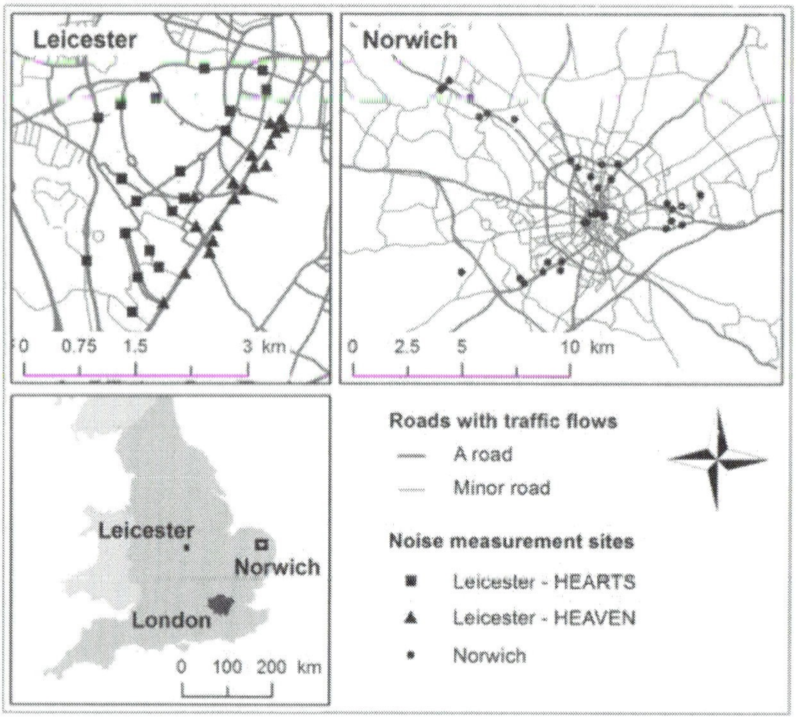

Figure 5. Location of noise measurement sites in Leicester and Norwich.

The focus of this exercise was assessing the model performance in terms of spatial contrast in noise levels. In Leicester a total of 38, 30-minute noise measurements were taken using Casella sound level meters (2xCEL480 and 2xCEL593.100a instruments). Measurements were taken at both facade and non-facade locations, with heights 3.5–4 m above ground level. All instruments were calibrated before and after each survey day using a Casella CEL calibrator (reference pressure: 94 dB) to ensure that the instrument had not drifted by more than 1 dB(A) over that day. Noise measurements were made in HEAVEN between the hours of 10:00 and 15:00 during August 2002 and in HEARTS between the hours of 09:00 and 16:30 during February 2005 (Goodman, 2005 and WHO, 2005).

In Norwich 35, 30-min noise level measurements were made using an Optimus CR:171B sound level meter, at 1.5 m above ground level, between the hours of 09:00 and 16:00, June 2014, next to residential properties. The microphone was placed at a height of 1.5 m to match the height of co-located air pollution monitoring. The noise sensor was calibrated at least three times per day using a CR:515 acoustic calibrator (reference pressure: 93.7 dB). All noise sensors had appropriate microphone shielding from the wind.

In Leicester we received traffic information from the Council's Airviro model (SMHI, 2014), used for local air quality management, which in turn received data from the cities' SCOOT (Split, Cycle, Offset, Optimisation Technique) (Imtech, 2013) system used for traffic control. In Norwich a version of the SATURN traffic model implemented by Norwich City Council was used to define composition, flows and speeds on main roads. Information from the traffic models and manual counts were used to produce time-varying information on flows, and in turn used to estimate noise levels for the different hours when noise measurements were made (e.g., 09:00 – 16:00). We used equivalent data to London on road geography, land cover and building heights in Leicester and Norwich to run TRANEX for model evaluation. Noise model estimates were made with respect to the height above ground levels of each instrument's microphone in Leicester (3.5–4 m) and Norwich (1.5 m). Noise level estimates from TRANEX were then compared with short-term noise measurements made in the cities of Leicester and Norwich.

A series of performance statistics were used to evaluate models including: Spearman's rank correlation coefficient (r), the coefficient of determination (r^2) (i.e. the sum of the squares of the errors divided by the sum of the squared differences between measurements and the mean of all measurements), root mean square error (RMSE), the variance of measured and modelled noise levels, and 95% confidence intervals (CI) for the regression fit between modelled and measured noise levels. We used Spearman's correlation to reflect the skewed nature of noise measurements and also because for exposure assessment we are interested in relative ranking of exposures as well as model error. These statistical tests were chosen to cover the key elements of characterising and assessing performance of environmental models as described in Bennett et al. (2013).

Exposure assessment

For each noise metric (e.g. $L_{Aeq,1hr}$), we assumed exposures to be equal to modelled noise levels. Road traffic noise exposures were calculated for all 189531 postcodes in London for each year between 2003 and 2010. Exposure assessment of the population of London was undertaken by assigning population headcounts to noise exposures calculated for each postcode location.

RESULTS

Model evaluation

Results of comparing measured and modelled noise levels are shown in Table 2 and Fig. 6. Table 2 shows summary statistics from the comparison of measured and modelled noise estimates in Leicester and Norwich. Fig. 6 shows scatterplots of measured against modelled noise estimates in Leicester and Norwich. A high level of correlation was seen between measured and modelled noise levels (Leicester: $r = 0.85$; Norwich: $r = 0.95$). The average error in predicted noise levels was 2.6 dB(A) for Leicester and 3.5 dB(A) for Norwich. In Leicester and Norwich 63% and 34% of sites, respectively, have predicted noise levels within ± 2 dB(A) of measured noise levels. In both cities, modelled noise levels tend to over-predict measured noise levels as indicated by regression fit lines in Fig. 6, but the model under-predicts the variability in measured noise levels as indicated by the values of variance of measured (Var_O) and modelled (Var_P) noise levels in Table 2. Pooling sites from Leicester and Norwich yields an $r = 0.90$ and an average model error of 3.1 dB(A). Overall the largest model errors relative to noise measurements are at sites where noise levels were measured <60 dB(A) and mostly are associated with low traffic flow in background areas.

Table 2. Summary statistics for the comparison of measured and modelled noise levels in Leicester and Norwich.

Location	N	Spearman's rho (r)	r^2	Var_O	Var_P	RMSE	Regression fit line		95% CI (lower, upper)
							β	Constant	
Leicester	38	0.95[a]	0.81[a]	35.5	26.4	2.6	1.09	−7.28	0.95, 1.22
Norwich	35	0.85[a]	0.72[a]	43	34.6	3.5	0.97	0.37	0.78, 1.17
All sites	73	0.90[a]	0.80[a]	45.9	37.3	3.1	1.02	−2.64	0.91, 1.12

Var_O is the variance of the measured values (i.e. observations).
Var_P is the variance of the modelled values (i.e. predictions).
a
$p = .000$.

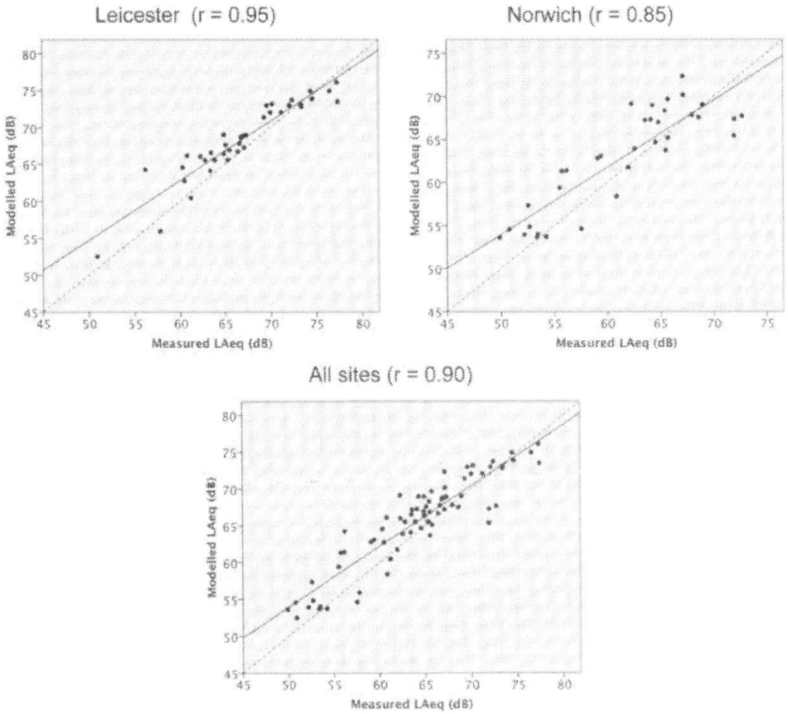

Fig. 6. Comparison of modelled and measured 30-min L_{Aeq} (dB) in Leicester, Norwich, and sites from both areas [regression fit line – solid line; 1-to-1 line – dashed line].

Modelled noise exposures in London

Table 3 shows summary statistics for modelled noise levels at 2008 postcode locations.Table 4 shows the number of postcode locations and number of people within (4 dB) bands of modelled noise exposures for $L_{Aeq,16hr}$ and L_{night}.

Table 3. Summary statistics for modelled noise levels (dB) at 2008 postcodes locations (n = 189521).

Noise metric	Min.	Max.	10th %ile	Median	90th %ile	Skewness
$L_{Aeq,16hr}$	54.8	83	54.9	55.6	68.3	1.39
L_{night}	49.2	78	49.3	50.1	63.5	1.35

Table 4. Cumulative modelled $L_{Aeq,16hr}$ and L_{night} road traffic noise exposures for postcodes and the resident population of London in 2008.

Noise level (dB)	$L_{Aeq,16hr}$ Cumulative counts of (% in brackets)		L_{night} Cumulative counts of (% in brackets)	
	Postcodes	People	Postcodes	People
48 – <51[a]	–	–	112207 (59.20)	6032157 (70.03)
51 – <54[a]	–	–	132,863 (70.10)	6,866969 (79.72)
54 – <57	120,224 (63.43)	6,399737 (74.30)	143,571 (75.75)	7,210561 (83.71)
57 – <60	136,596 (72.07)	6,984673 (81.09)	155,100 (81.83)	7,595073 (88.18)
60 – <63	146,909 (77.51)	7,317983 (84.96)	168,361 (88.83)	8,043448 (93.38)
63 – <66	159,126 (83.96)	7,727753 (89.72)	180,115 (95.03)	8,407451 (97.61)
66 – <69	173,588 (91.59)	8,221195 (95.45)	186,669 (98.49)	8,571285 (99.51)
69 – <72	183,723 (96.94)	8,508549 (98.78)	188,831 (99.63)	8,607192 (99.93)
72 – <75	187,834 (99.10)	8,593616 (99.77)	189434 (99.95)	8,612624 (99.99)
75 – <78	189,112 (99.78)	8,610546 (99.97)	189530 (100)	8,613482 (100)
78 – <81	189,510 (99.99)	8,613230 (100)	189,531 (100)	8,613526 (100)
81 – <84[b]	189,531 (100)	8,613526 (100)	–	–

[a]Below the minimum (see Table 3) modelled noise level for $L_{Aeq,16hr}$.

[b]Above the maximum (see Table 3) modelled noise level for L_{night}.

For $L_{Aeq,16hr}$ there is < 1 dB (i.e. median – min) variability across 50% of postcode locations (Table 3); 10% of the population have $L_{Aeq,16hr}$ and L_{night} noise levels > 68.3 dB and >63.5 dB, respectively. For $L_{Aeq,16hr}$ and L_{night}, 74% and 70%, respectively, of the population have modelled road traffic noise exposures in the lowest 4 dB category of noise exposures (as shown in Table 4). Approximately 19% and 12% of the population are exposed to road traffic $L_{Aeq,16hr}$ noise levels ≥60 dB and ≥65 dB, respectively. Approximately 19% and 12% of the population are exposed to road traffic L_{night} noise levels ≥55 dB and ≥60 dB, respectively.

TRANEX does not produce a continuous surface of noise level estimates (i.e. regular grid), but Fig. 7 shows the spatial variability in modelled road traffic noise exposures at postcodes locations within London for 2008.

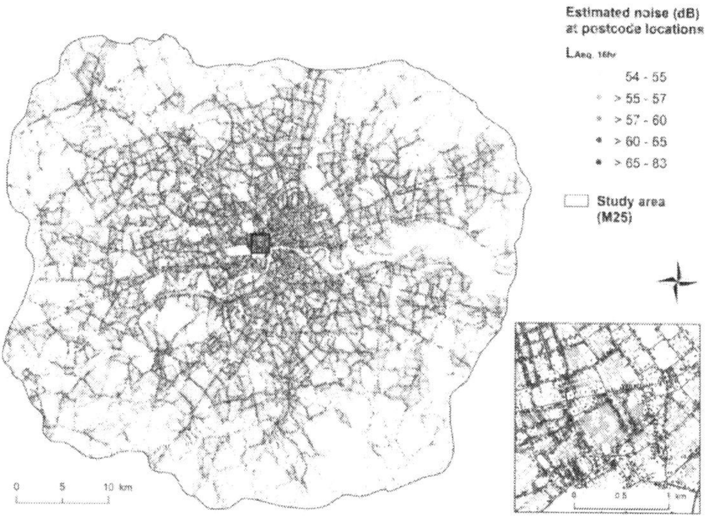

Figure 7. Modelled $L_{Aeq,16hr}$ (dB) noise exposures at postcode locations in London.

Changes in modelled noise exposures over time (2003–2010)

Differences over the study period (2003–2010) in noise exposures were calculated by subtracting values of $L_{Aeq,16hr}$ and L_{night} predicted at postcode locations (n = 189531) for 2003 from values predicted for 2010 (Table 5). Relatively large negative and positive changes were seen at a small number of postcodes. This may be due to traffic interventions (road closures, road building, traffic diversions etc.) between 2003 and 2010 (N.B. the comparison is for the same postcode locations). The average change for $L_{Aeq,16hr}$ and L_{night} is < 0.3 dB. Traffic flows over this period have generally increased but lower speeds due to traffic congestion may counteract the ability of traffic flows to raise noise levels. Approximately 54% of postcodes for both $L_{Aeq,16hr}$ and L_{night} have an increase in modelled noise exposures between 2003 and 2010. The majority of postcode locations (96%) have changes in predicted $L_{Aeq,16hr}$ and L_{night} noise exposures of < ±2 dB between 2003 and 2010.

Table 5. Changes in predicted noise levels (dB) at postcode locations: subtracting 2003 values of $L_{Aeq,16hr}$ and L_{night} from 2010 values.

Noise metric	Mean	Standard deviation	1st %ile	99th %ile	Minimum	Maximum
$L_{Aeq,16hr}$	0.25	0.89	−0.34	2.91	−12.52	21.91
L_{night}	0.26	0.97	−0.36	2.94	−12.42	21.56

DISCUSSION

We developed a traffic noise model (TRANEX) based on the CoRTN method for population exposure assessment in London. The model was developed in open-source GIS (via R). TRANEX was evaluated against short-term noise measurements made in Leicester and Norwich, UK. We undertook population exposure assessment using postcode locations in London for the period 2003 – 2010.

Performance of TRANEX

To date, we have not undertaken an evaluation study in London (i.e. where TRANEX was implemented for exposure assessment). The areas in Leicester and Norwich where noise measurements were made, however, have similar characteristics to many areas in London in terms of spatial contrasts in heavily trafficked roads, street canyons and open space. The study areas in Leicester and Norwich, for example, include several roads with >20,000 AADT. One advantage of the noise measurements made in Leicester and Norwich for evaluating TRANEX is that they were not influenced by noise from aircraft and railways. Taking all sites, the average error for our model was 3.1 dB(A) and the Spearman's correlation was 0.90 (p = .000) Comparing modelled and measured hourly L_{Aeq} is a stringent test of models; it is likely that model errors would be lower if compared with measurements over longer averaging times ($L_{Aeq,16hr}$, L_{night}). The better performance of TRANEX in Leicester compared to Norwich may be in part due to generally more representative traffic information in Leicester.

Noise model evaluations in the open literature, comparing estimated noise levels with noise measurements, are relatively scarce, and like the measurements presented in this study tend, by necessity, to be for short time periods when the focus is on assessing spatial contrasts in the performance of noise models. In Vancouver, Canada, Gan et al. (2012) compared measured (LDL 870 Environmental Noise Analyser) 5-min daytime (08:00–18:00) A-weighted noise levels with noise level predictions from CadnaA (using RLS-90; the standard method used in Germany) for 103 roadside locations. On average, measured 5-min noise levels were 6.3 dB higher than L_{day} but showed a good level of correlation (r = 0.62). The underestimation in 5-min noise levels may be due in part to the different averaging periods for measured and modelled noise levels. Ko et al. (2011)compared modelled noise levels from the RLS-90 method in SoundPLAN 6.5 with daytime and night-time noise measurements made at 25 locations in Chungju, Korea (N.B. no details were given of the measurement instruments). Measured noise levels were in the range 65–77 dB. Overall the average error in predicted noise levels was 0.6 dB with an r^2 of 0.50. Mioduszewski et al. (2011) compared noise predictions from CadnaA with noise measurements at 25 locations for L_{day}, L_{eve} and L_{night}. Differences between measured and modelled noise levels were in the range −3.1–8.0 dB; for $L_{day}$and L_{night} the range was −3.1 to 4.4. dB (average = 0.1 dB) and 0.8–8.0 dB (average = 4.7 dB), respectively. In Santiago, Chile, Suárez and Barros (2014) used CadnaA, applying different model options (RLS-90, the Swiss model STL96, the Nordic method SP48, and CoRTN) to predict L_{Aeq} noise levels for comparison with 15-min noise measurements made during the day (07:00 – 21:00) at 52 (roadside, suburban and rural) locations. Over 50% of differences between measured and modelled noise levels were less than 1 dB, with models tending to under-predict noise measurements. Errors in predicted noise levels were on average lower close to main roads (1.7 dB) than on local roads (3.1 dB). Lee et al. (2014) compared measured and modelled noise levels in three cities in the United States. Noise measurements (10-min L_{eq}) were collected between 09:00 and 17:00 h during weekdays using a 3 M SD-200 sound level meter in Atlanta (20 sites), Los Angeles (26 sites) and New York City (26 sites). Noise levels were estimated, for each of the 72 sites for the corresponding hour of the day when noise measurements were made, using the US Federal Highways Agency Traffic Noise Model (TNM). Noise levels were predicted using local traffic models in each city and then again from 10-min observed traffic counts made at the same time as noise measurements. TNM using traffic counts explained a high proportion of the variability in measured 10-min noise levels (R^2: 0.56–0.73) but model performance using modelled long-term traffic flows was overall poor (R^2: 0.08–0.42). TNM tended to under-estimate noise measurements, which may be because only the streets where noise measurements took place were included in the model. Based on comparison of our model evaluation with other studies we concluded that the model performance in our study is acceptable. Most importantly for exposure assessment, where relative ranking of exposures is important, model evaluation in this study showed a high level of (rank) correlation between noise levels estimates and noise measurements (r: 0.85–0.95).

Modelled noise exposures in London

According to the World Health Organisation (WHO) community noise guidelines (2009), average outdoor noise levels in residential areas should not exceed 55 dB and 40 dB for $L_{Aeq,16hr}$ and L_{night}. We modelled exposures of the resident population in London using 189531 postcodes locations with a population of 8.61 million and presented results for 2008 as an example of our model output. For modelled $L_{Aeq,16hr}$ and L_{night}, 19% and 100%, respectively, of the population have noise levels exceeding the WHO guidelines. The CoRTN method implemented in TRANEX predicts a minimum night-time $L_{Aeq,1hr}$ of 38.0 dB for a single minor road. The minimum modelled value of night-time $L_{Aeq,1hr}$ in London was 42.4 dB because we combine the contribution of all sources up to 0.5 km or 1 km.

WHO also state that 20% of European Union populations are exposed to daytime levels exceeding 65 dB and 30% are exposed to night-time levels greater than 55 dB. For London, our results gave 12% (~1.03 million people) and 19% (~1.6 million people) of the population exposed to noise levels above these thresholds, respectively. Our estimates are based on road traffic noise alone and do not include incremental noise levels from aircraft and railways applicable to some of the resident population. A substantial number of people living in London are thus exposed to noise levels that are unacceptable during daytime and at night.

Strengths and limitations

We have implemented the main aspects of the CoRTN in TRANEX. Our model does, however, have limitations. We have not included elevated road sections (e.g. flyovers, bridges). There is no information included in TRANEX on existing noise barriers. We have modelled noise propagation on a flat world – although London is relatively low-lying there are suburban areas with undulating terrain that will affect the propagation of noise levels between sources and receptors. We assumed minor roads to have a constant traffic flow of 600 vehicles day^{-1}. Notwithstanding that our results showed this to be a good approximation for noise levels on roads with low flows (i.e. ~50 vehicles hour^{-1}during the daytime), we may have under-estimated noise levels for some minor roads with higher flows (i.e. those roads that were not part in the LAEI). The LAEI does, however, include roads with relatively low flows (i.e. < 1000 vehicles day^{-1}). Furthermore, we have applied the CoRTN method with a single value of average vehicle speed for each road link because at the time of this study we were limited to this information. Thus, our model is not 'dynamic' in terms of representing varying vehicle speeds along road links. Others (Guarnaccia, 2013 and Iannone et al., 2013) have shown that noise levels are different around road intersections (acceleration/deceleration) than at mid-block (i.e. the free-flowing section of a road link) for the same volume of traffic, which means that we may have variable errors in modelled noise levels associated with changes in vehicle speeds along road links relative the average vehicle speeds that we have applied.

We have calculated noise levels at a single façade that is closest to the nearest main road for each dwelling. We decided to model noise at one point per building due to time constraints in modelling such a large number of locations. Thus, we may have misclassified exposures for other façades (i.e. rooms) within dwellings. Receptor placement is a pre-process to noise modelling and in a separate model script, which does not prohibit future applications by us, or others, in having a more detailed definition of receptors. An alternative method, for example, is to estimate noise levels at several points around each building and from them calculate the average (or median) noise level. This would, however, substantially increase the processing time for our model to estimate noise levels across London, and this was not feasible in the lifetime of this study. Noise estimates are universally made at 4 m above ground and this will misclassify noise levels to a varying degree for dwellings in high-rise accommodation. We were unable to include the gradient correction for uphill flow on single carriageways. Tunnel elements being excluded may lead to under-prediction of noise in the areas adjacent to tunnel mouths due to lack of propagation of retro-reflected sound from tunnel walls. As the CoRTN method produces estimates of noise levels in values of L_{10} we used the method by Abbott and Nelson (2002) to convert L_{10} to L_{eq}. Abbott and Nelson (2002)showed that there was high correlation ($r^2 = 0.90$) and relatively a small standard error of 2.1 dB (from comparison of 460 measurements of $L_{eq,1hr}$ and predicted values of $L_{10,1hr}$in free-flowing traffic (i.e. as per the traffic model used in this study). It is expected that the magnitude of errors would be reduced for longer averaging periods such as those used in this study ($L_{Aeq,16hr}$; L_{night}). Indeed, Abbott and Nelson (2002) also showed that for 18-h averaging periods (06:00 – 24:00) comparison of measurements (1024 measurements at 76 sites) of $L_{eq,18hr}$ and predicted values of $L_{eq,18hr}$ yielded a R^2 of 0.97 and the standard error was 0.85 dB. Finally, we estimated that 100% of people in London live in locations where the WHO community noise guideline of 40 dB for L_{night} is exceeded. The CoRTN method is, however, limited to a minimum noise value for $L_{Aeq,1hr}$ of 38.0 dB during the quietest part of the night-time, which, with the data supplied to our model, resulted in a minimum value of 42.4 dB. By using the CoRTN method, we may have over-estimated night-time noise especially for people living in the quietest areas.

Despite these limitations, our model has a number of strengths. In particular, detailed information on traffic for ~63000 road links including varying flows and speeds for each year in the study period (2003–2010), and detailed information on land cover and heights of individual buildings within London. Through the implementation of the model in PostgreSQL, TRANEX offers excellent efficiency in terms of processing time for a large number of sources compared to other standard software platforms such as ArcGIS (~11 days processing time for 189531 postcodes on a 3.40 Ghz, 16 Gb RAM, 64 Bit, Intel i7-3770). TRANEX has been developed so that it is transferable to other cities. In the UK this means that it can be applied in most areas as long as there is sufficiently detailed information on traffic flows, composition and speeds.

Future work

In consideration of other sources of environmental noise in the epidemiological studies, modelled values of noise levels from air and railways (not reported here) were provided by Department of Environment, Food and Rural Affairs (DEFRA) via permissions obtained from the rail industry and Civil Aviation Authority (CAA). By mid-2014 we had commenced running TRANEX for all ~3 million address point locations in London. Following the discussion on limitations of our current model, our aim is to make a number of improvements in future applications. We aim, for example, to develop a method to improve noise exposure estimation for dwellings on minor roads. We are currently considering a method that combines information on the order of connectivity between main roads and minor roads; e.g. a minor road connecting two main roads (i.e. 1st order) will likely have higher traffic flows than a minor road several orders "upstream" (e.g. dead-end roads/cul-de-sacs). This work would involve undertaking turning surveys between main roads and minor roads, and counting traffic along minor roads of different orders of connectivity to main roads. The work would also require a comprehensive series of noise measurements along different types of minor road to evaluate the approach. We also aim to improve estimation for address locations in high-rise accommodation by including terms in the model to calculate noise levels at receptors >4 m above ground level. The latter will depend on our ability to characterise the height of vertical address points.

Another area for improvement is to introduce 'dynamic' speed profiles on road links to better represent the acceleration/deceleration and free-flow phases of traffic. This would involve improvements to the traffic model that we have used in this study. Indeed, our model implementation using a series of points spaced along road links is ideal for applying variable speed information if it becomes available. We also intend to look at including an "intelligent", variable-sized buffer for selecting noise sources; for example, a fixed buffer size of 500 m may be too large when receptors are close to one or more main roads in a densely built up area, particularly in continuous street canyons (i.e. in this situation, noise from roads several hundred metres away will make negligible difference to incremental noise levels); for example, a 1000 m buffer may even be too small when a receptor is downwind of a major highway in open, flat terrain. We will also consider including atmospheric absorption and refraction and meteorological effects that are accounted for in other models (e.g. RLS-90, Nord 2000, HARMONOISE) and use wind direction to set the variable buffer for selecting noise sources. We intend, or invite others using our freely available software, to compare exposure estimates from our model with other models including those that are commercially available (e.g. SoundPlan, CadnaA etc.). We will compare noise levels from TRANEX with those from our implementation of the CNOSSOS-EU method (Kephalopoulos et al., 2014) (i.e. the standard method being adopted in the EU for noise assessments), which is currently in development, for different times of the day. Finally, we will undertake additional evaluation studies and make continuous noise measurements over longer periods (e.g. 24-hours) to allow comparison between noise measurements and the main noise metrics produced by TRANEX ($L_{Aeq,16hr}$, L_{night}).

ACKNOWLEDGEMENTS

We are grateful to Stephen Turner (Department for Environment, Food & Rural Affairs) as advisor on environmental noise in the TRAFFIC project. We thank Rob Mills and Bradley Lewis at Norwich City Council for provision of data from the Council's traffic model and data on manual traffic counts made within the study area. We thank Leicester City Council for modelled traffic information from their Airvrio and SCOOT systems for the Leicester Study area. Funding for this study came from the NERC (Environmental Exposures and Health Initiative: EEHI) funded 'Traffic and Health in London: TRAFFIC study' (grant number: NE/I00789X/1). The data described in this paper were used under licence from the EDINA Digimap® (http://digimap.edina.ac.uk/digimap/home) as part of the agreement for use of Ordnance Survey® data in Higher Education establishments in the UK. Information on traffic was made available by King's College London and is equivalent to the data held in the LAEI (2010).

REFERENCES

1. Abbott, P.G., Nelson, P.M., 2002. Converting the UK Traffic Noise Index LA10,18h to EU Noise Indices for Noise Mapping PR/SE/451/02 [EPG 1/2/37]. TRL Report for DEFRA.
2. Babisch, W., van Kamp, I., 2009. Exposure-response relationship of the association between aircraft noise and the risk of hypertension. Noise Health 11, 161e168.
3. Babisch, W., Pershagen, G., Selander, J., Houthuijs, D., Breugelmans, O., Cadum, E., et al., 2013. Noise annoyance e a modifier of the association between noise level and cardiovascular health? Sci. Tot. Environ. 452e453, 50e57.
4. Beevers, S., Carslaw, D., Westmoreland, E., Mittal, H., 2009. Air Pollution and Emissions Trends in London. http://uk-air.defra.gov.uk/assets/documents/reports/cat05/1004010934_MeasurementvsEmissionsTrends.pdf (accessed 28.07.14.).
5. Bennett, N.D., Croke, B.F.W., Guariso, G., Guillaume, J.H.A., Hamilton, S.H., Jakeman, A.J., Marsili-Libelli, S., Newham, L.T.H., Norton, J.P., Perrin, C., Pierce, S.A., Robson, B., Seppelt, R., Voinov, A.A., Fath, B.D., Andreassian, V., 2013. Characterising performance of environmental models. Environ. Mod. Soft. 40, 1e20.
6. Clark, C., Crombie, R., Head, J., van Kamp, I., van Kempen, E., Stansfeld, S.A., 2012. Does traffic-related air pollution explain associations of aircraft and road traffic noise exposure on children's health and cognition? A secondary analysis of the United Kingdom sample from the RANCH project. Am. J. Epidemiol. 176, 327e337.
7. Clark, C., Head, J., Stansfeld, S.A., 2013. Longitudinal effects of aircraft noise exposure on children's health and cognition: a six-year follow-up of the UK RANCH cohort. J. Environ. Psychol. 35, 1e9.
8. DEFRA, 2008. Noise Mapping England. Online Resource. http://services.defra.gov. uk/wps/portal/noise (accessed 31.07.14.).
9. Department of Transport, 1988. Calculation of Road Traffic Noise. HMSO.
10. Evans, G., Hygge, S., 2007. Noise and performance in adults and children. In: Luxon, L., Prasher, D. (Eds.), Noise and its Effects. Whurr Publishers, London.
11. Gan, W.Q., McLean, K., Brauer, M., Chiarello, S.A., Davies, H.W., 2012. Modeling population exposure to community noise and air pollution in a large metropolitan area. Environ. Res. 116, 11e16.

12. Goodman, P.S., 2005. HEARTS (Health Effects and Risks of Transport Systems): Report on Noise Modelling of the HEARTS Study Area in Leicester Using AVTUNE. Draft Internal Report of the HEARTS Project, EU 5th Framework. HEA- 280705eITS03. Institute for Transport Studies, University of Leeds.
13. Guarnaccia, C., 2013. Advanced tools for traffic noise modelling and prediction. WSEAS Trans. Syst. 2 (12), 121e130. Print ISSN: 1109-2777, e-ISSN: 2224-2678.
14. Hammer, M.S., Swinburn, T.K., Neitzel, R.L., 2014. Environmental noise pollution in the United States: developing an effective public health response. Environ. Health Perspect. 122 (02), 115e119.
15. Hansell, A.L., Blangiardo, M., Fortunato, L., Floud, S., de Hoogh, K., Fecht, D., et al., 2013. Aircraft noise and cardiovascular disease near London Heathrow Airport. Br. Med. J. 8, 347:f5432.
16. HMSO, 2006. Environmental Noise (England) Regulations 2006. SI 2006 No. 2238. Imtech, 2013. SCOOT e the World's Leading Adaptive Traffic Control System. Online Resource. http://www.scoot-utc.co.uk/ (accessed 31.07.14.).
17. Iannone, G., Guarnaccia, C., Quartieri, J., 2013. Speed distribution influence in traffic acoustical noise prediction. Environ. Eng. Manage. J. 12 (3), 493e501.
18. Kephalopoulos, S., Paviotti, M., Anfosso-Led ee, F., Van Maercke, D., Shilton, S., Jones, N., 2014. Advances in the development of common noise assessment methods in Europe: the CNOSSOS-EU framework for strategic environmental noise mapping. Sci. Tot. Environ. 482e483, 400e410.
19. Ko, J.E., Chang II, S., Lee, B.J., 2011. Noise impact assessment by utilizing noise map and GIS: a case study in the city of Chungju, Republic of Korea. Appl. Acoust. 72, 550e554.
20. Lee, E.Y., Jerrett, M., Ross, Z., Coogan, P.F., Seto, E.Y.W., 2014. Assessment of trafficrelated noise in three cities in the United States. Environ. Res. 132, 182e189.
21. London Atmospheric Emissions Inventory (LAEI): http://data.london.gov.uk/ datastore/package/london-atmospheric-emissions-inventory-2010 (accessed on 28.07.14.).
22. Mioduszewski, P., Ejsmont, J.A., Grabowski, J., Karpinski, D., 2011. Noise map vali- dation by continuous noise monitoring. Appl. Acoust. 72, 582e589.
23. Pamanikabud, P., Tansatcha, M., 2003. Geographical information system for noise analysis and forecasting with the appearance of barriers. Environ. Mod. Soft 18, 959e973.
24. SMHI. Airviro. Swedish Meteorological and Hydrological Institute. Online resource: http://www.smhi.se/airviro, 2014; (accessed 31.07.14.).

25. Sorensen, M., Andersen, Z.J., Nordsborg, R.B., Jensen, S.S., Lillelund, K.G., Beelen, R., et al., 2012. Road traffic noise and incident myocardial infarction: a prospective cohort study. PLoS One 7 (6), e39283. http://dx.doi.org/10.1371/ journal.pone.0039283.
26. Stansfeld, S.A., Matheson, M.P., 2003. Noise pollution: non-auditory effects on health. Br. Med. Bull. 68, 243e257.
27. Suarez, E., Barros, J.L., 2014. Traf fic noise mapping of the city of Santiago de Chile. Sci. Tot Environ. 466e467, 539e546.
28. World Health Organisation (WHO), 2005. Health Effects of Transport-related Air Pollution. WHO Publ., Copenhagen. http://www.euro.who.int/__data/assets/ pdf_file/0013/91102/E88772.pdf (accessed on 28.07.14.).
29. World Health Organisation, 2009. Night Noise Guidelines for Europe. Denmark, Copenhagen. Available at: http://www.euro.who.int/__data/assets/pdf_file/ 0017/43316/E92845.pdf (accessed 29.07.14.).
30. World Health Organisation and Joint Research Council (WHOeJRC), 2011. Burden of Disease from Environmental Noise. Quantification of Health Life Years Lost in Europe. Denmark, Copenhagen. Available at: http://www.euro.who.int/__data/ assets/pdf_file/0008/136466/e94888.pdf (accessed 29.07. 14.).

CITATION

John Gulliver, David Morley, Danielle Vienneau, Federico Fabbri, Margaret Bell, Paul Goodman, Sean Beevers, David Dajnak, Frank J Kelly, Daniela Fecht, Development of an open-source road traffic noise model for exposure assessment, Environmental Modelling & Software, Available online 8 January 2015, ISSN 1364-8152, http://dx.doi.org/10.1016/j.envsoft.2014.12.022.

CHAPTER 5

External Costs to Parties Involved in Highway Traffic Accidents: The Perspective of Highway Users

Rong-Chang Jou and Tzu-Ying Chen

Department of Civil Engineering, National Chi Nan University, No.1, University Rd, Puli, Nantou County 54561, Taiwan

ABSTRACT

This study explores highway travellers' willingness to pay (WTP) for external costs caused by traffic accidents. There are a number of further external costs, the paper focuses on two externalities: air pollution and time delays. Data collection was performed using the face-to-face survey method, and the surveys were carried out at highway rest areas. Air pollution and time delays were divided into three levels of severity (light, moderate and severe) to obtain the interviewees' WTP according to each level of severity. The result of this study demonstrates that there are many samples with zero WTP because penalties for pollution caused by traffic accidents are not currently enforced in Taiwan. Thus, the spike model was adopted in this study to overcome any estimation error that might be caused by excessive NT$0 WTP samples. The results show that variables such as age, education, income and willingness to participate in activities of environmental protection have a positive effect on WTP for air pollution, whereas variables such as occupation, travel purpose, traveller identity, travel time and travel distance have a significantly positive effect on WTP for time delays. WTP for nitrogen dioxide (NO_2) is NT$8862–11,502/metric ton (US$1 = NT$30) and WTP for carbon dioxide (CO_2) is NT$1070–2693/metric ton. Moreover, WTP for time delays is NT$960–1320/h. The findings of this study not only demonstrate WTP for air pollution and time delays in the minds of parties to traffic accidents but also help to provide agencies with a basis to formulate applicable penalties in the future.

INTRODUCTION

Like other developing countries in Asia, traffic accidents in Taiwan have been increasing at an alarming rate in the recent years. Highway traffic accidents cause not only internal costs directly (*i.e.*, loss of lives and property damage) but also external costs indirectly (*i.e.*, travel time delay, energy consuming and air pollution emission). The internal costs of highway traffic accidents are introduced in our previous study [1], so that this paper focuses on analysing the external costs of highway traffic accidents. Three types of external cost are identified: system externalities, physical injury externalities, and traffic volume externalities. System externalities costs that road users impose on the rest of society, are estimated to be about 30% of the total cost of traffic injury in Norway [2]. Forkenbrock [3] estimates that the external costs per ton-mile of freight shipped by truck include accidents (fatalities, injuries, and property damage), emissions (air pollution and greenhouse gases), and noise. Zegras [4] indicates that personal motor vehicles are the type of transportation that has the highest cost of travel, especially passenger cars, compared to other modes, which account for 46.7% and 55.3% of the total cost in traffic accidents and air pollution, respectively. Shafie-Pour and Ardestani [5] point out that the major cause of air pollution in Iran's capital is the emissions of 2.4 million motor vehicles.

According to statistical research conducted by the International Energy Agency in 2004, transportation is responsible for one-quarter of the world's energy consumption. In Taiwan, there are many automobiles and motorcycles that have high mobility. As vehicles travel on the roads, carbon dioxide is generated by burning fossil fuels, raising the average global temperature. This carbon dioxide emission is a major cause of global warming and climate change, which will harm the global ecological environment as the situation worsens so severely that it may even threaten humanity. The major pollutants from running and idle vehicles' emissions are carbon monoxide, nitrogen oxide and carbon dioxide. These pollutants cause deterioration in air quality and cause damage to human health and have gradually gained attention from many countries. However, in addition to air pollution and time delays caused by regular transportation activities, when traffic accidents occur, the parties involved create extra air pollution and time delays, imperceptibly increasing social costs. To determine an effective measure to decrease air pollution caused by motor vehicles, aside from creating alternative, energy-efficient vehicles, we must understand how to restrict vehicle usage from the perspective of pollution economics. In this article, we estimate two external costs: air pollution and time delays.

Little research has focused on the air pollution caused by car accidents and the time delays inflicted on those affected by accidents, as most research has examined regional air pollution or time delays. There is no international, standardised measurement of the cost of air pollution. Some methods have employed to measure the cost of air pollution: one method has estimated people's willingness to pay (WTP) based on determining total air pollution emissions or the concentration thereof [6,7,8,9,10]. Another method calculates people's WTP for reducing the health damage caused by air pollution [5,11,12,13]. Since the 1998 implementation of air pollution control fees in Taiwan, fee collection has been limited to stationary sources of air pollution such as sulphur oxide and nitric oxide; for non-traditional pollutants such as carbon dioxide, only a vehicle refuelling tax has been adopted, and no common tax collection standards have been established. Therefore, the economic value of air pollution cannot yet be determined by measuring the total amount of pollution emissions. On the other hand, when car accidents occur, time uncertainty and extra time costs are imposed on travellers. Furthermore, traffic delays also lead to congestion. In recent years, research has proven not only that travellers are willing to pay money to save time [14,15,16,17,18] but also that WTP estimated by the stated preference method is relatively accurate.

However, because the travel time delay costs due to traffic congestion cannot be the responsibility of specific road user, the economic value of congestion cost for affected road user cannot be estimated directly. Therefore, this paper estimates the traffic accident parties' WTP of external costs (travel time delay, air pollution emission) caused by traffic accidents. These results can be employed for establishing a fine standard. It should be noted that improving environmental quality comes after income considerations. In addition, the calculation of WTP environmental costs is based on a fictional market; comparisons and market conditions were not accessible to interviewees, which may have resulted in extremely wild guesses. In traditional research on environmental issues, the Contingent Valuation Method (CVM) has commonly been adopted to determine interviewees' WTP. CVM was first introduced by Ciriacy-Wantrup [19], who measured the value of environmental resources through interview surveys. Since then, CVM has been broadly applied in the measurement of non-market goods. For example, Thunberg and Shabman [20] have adopted the CVM method on the costs of floods that landowners are willing to pay; Wang and Mullahy [12] have utilised CVM to determine the amount that residents of Chongqing were willing to pay to reduce air pollution damages; in Spain, Lera-López *et al.* [9] have performed a regional study on five of the major roads between Spanish Navarre and the Pyrenees and have obtained residents' WTP to reduce air pollution and noise. Therefore, the present study adopted the CVM, combining the dichotomous choice and open-inquiry method to allow interviewees to express their WTP based on their experience or knowledge, so as to obtain a more effective WTP.

However, according to Davies and Mazurek [21], once environmental-related activities affect the daily lives of interviewees, a large amount of zero samples appear. When a large portion of the data is zero, the sample's data distribution clearly does not fall within a normal distribution. If multiple regression or binary probability models (*i.e.*, logit or probit bivariate models) are used under this circumstance, model estimation error may occur.

In recent years, there have been many methods for handling data with large amounts of zero samples. Most methods have adopted the limited dependent variables model that is capable of analysing censored data, among which the Tobit model, developed by Tobin [22], is the most widely used. The Tobit model supposes that all interviewees are willing to participate and pay, treating the zero value as corner solutions. The model is specified for the analysis of repeated dichotomous choice data and produces only positive estimates of WTP, and it has been extensively used in studies related to economics, the social sciences, the environment, medicine and traffic accident rates [23,24,25,26]. The second approach is to simply remove these observations from the sample [27,28], however, such a complete removal may not be appropriate when mean bids are calculated [29]. Reichl and Frühwirth-Schnatter [30] address the problem of negative estimates of willingness to pay and find that there exist a number of goods and services, especially in the fields of marketing and environmental valuation, for which only zero or positive WTP is meaningful. Their model restricts the domain of the estimates of WTP to strictly positive values, while also allowing for the detection of zero WTP. The model is tested on a simulated and a real data set.

The third approach explicitly allows for a point mass at zero, which is the truncation at zero of the preferred WTP's distribution. In 1997, Kriström [31] proposed the use of the spike model to calculate WTP. This model uses the random utility theory as its foundation, allowing for data that indicates zero WTP while avoiding the irrational situation in which WTP is a negative value. The differences between the spike model and the Tobit model are the following: (1) the spike model obtains the WTP price range from the triple-bound dichotomous choice (TBDC) scenario, which is different from the Tobit model that obtains the true WTP; and (2) the Tobit model assumes that the error term is normally distributed, whereas the spike model follows the Gumbel distribution. Since the present study applies TBDC scenarios instead of asking the true WTP, the spike model was chosen. Due to the capability of handling data with a high proportion of zero values (e.g., above 10% of the total sample), many studies gradually have begun to utilise the spike model to discuss WTP for environmental pollution [9,32,33,34,35]. Those studies have shown that the spike model reflects interviewees' WTP more realistically. Thus, in the hope of establishing a logical economic model, the spike model is applied in this study.

The structure of this study is as follows. The first section describes the study's research motivation and objective, including a literature review. The second section contains the construction of the model. The third section explains the research data and analysis. The fourth section shows the results of the model calculation. Finally, the fifth section contains conclusions and recommendations.

METHODS

Because sanctions for carbon emissions caused by traffic accidents have not been implemented, the economic cost of such measures to the parties involved cannot be evaluated. Thus, this study utilised the CVM to obtain interviewees' subjective valuation of compensation for carbon emissions caused by traffic accidents; subsequently, the acceptance of this non-market good with an equivalent value was evaluated.

Following Hanemann [36] by assuming that the utility function consists of observable and unobservable components, the individual's utility function can be written as

$$U_i = V_i + \varepsilon_i \tag{1}$$

When parties to traffic accidents are willing to pay the amount A specified in the scenario, a new utility$V1$ becomes the utility function that the respondent is willing to pay for air pollution (or time delays), which is higher than the original utility (V_0). Therefore, the utility function can be rewritten as Equation (2):

$$V_1(Y - A, X, Q_1) + \varepsilon_1 \geq V_0(Y, X, Q_0) + \varepsilon_0 \tag{2}$$

where Y is income or the available assets, X is a vector of socioeconomic characteristics, Q is a vector of variables other than X, such as the awareness of environmental effects *etc.*, and $\varepsilon 0$ and $\varepsilon 1$ are random terms with an independently and identically (iid) Gumbel distribution (zero means). We assume that the CV survey creating a hypothetical market leads to a change in Q from Q_0 *to* Q_1. In this case, the interviewee prefers the new utility and is willing to pay the amount A, that is, his or her income or available assets will be reduced by A (Y1=Y−A). Thus, the probability function that a given individual pays the amount in the new state can be derived as follows:

$$
\begin{aligned}
&\Pr(Yes) = \Pr(\Delta V(.) \geq \varepsilon) = F_\varepsilon(\Delta V(.)) \\
&\Delta V = V_1(Y - A, X, Q_1) - V_0(Y, X, Q_0) \\
&\varepsilon = \varepsilon_0 - \varepsilon_1
\end{aligned}
\quad (3)
$$

When the interviewee's WTP for air pollution or time delays is greater than the bid offered in the scenario (A), he or she will be willing to pay that amount. The probability of the individual paying the amount A in the new state can thus be derived as

$$
\Pr(Yes) = \Pr(WTP \geq A) = 1 - F_{WTP}(A) = F_\varepsilon(\Delta V(.)) \quad (4)
$$

Equation (3) can be further rewritten as Equation (5) based on the assumption regarding the form of the functions V_1 and V_0:

$$
\begin{aligned}
\Delta V(.) &= \alpha_1 + \beta(Y - A) + \delta Q_1 - (\alpha_0 \mid \beta Y + \delta Q_0) + \gamma X \\
&= \alpha_1 - \alpha_0 - \beta A + \gamma X + \delta(Q_1 - Q_0) \\
&= \alpha - \beta A + \gamma X + \delta Q
\end{aligned}
\quad (5)
$$

Since ε is logistically distributed, its cumulative distribution function can be expressed as [31]:

$$
F_\varepsilon(\Delta V(.)) = \frac{1}{1 + \exp[-\alpha + \beta A - \gamma X - \delta Q]} \quad (6)
$$

Assuming that the random utility function of personal decisions are as described in Equations (1)–(4), and that the range of price A can be zero or positive; therefore, the function FWTP(A) is defined as follows:

$$F_{WTP}(A) = \begin{cases} P, & A = 0 \\ G_{WTP}(A), & A > 0 \end{cases} \tag{7}$$

where p belongs to (0, 1) and FWTP(A) is a continuous increasing function, whereas GWTP(0)=p and limA→∞GWTP(A)=1. Thus, P is greater than zero and manifests as a discontinuous spike in the graphic. *P*is also the probability that a respondent's preferred fine is zero, which can be obtained from model estimation results. Furthermore, Equation (7) can be rewritten as Equation (8):

$$F_{WTP}(A) = \begin{cases} [1 + \exp(\alpha + \gamma X + \delta Q)]^{-1}, & A = 0 \\ [1 + \exp(\alpha - \beta A + \gamma X + \delta Q)]^{-1}, & A > 0 \end{cases} \tag{8}$$

The expectation value of WTP by the interviewee, E(WTP) is shown by Equation (9):

$$\begin{aligned} E(WTP) &= \int_0^\infty 1 - F_{WTP}(A)dA = \int_0^\infty \left(\frac{e^{\alpha-\beta A+\gamma X+\delta Q}}{1+e^{\alpha-\beta A+\gamma X+\delta Q}}\right) dA \\ &= \frac{-1}{\beta}\left\{\lim_{A->\infty}[\ln(1+e^{\alpha-\beta A+\gamma X+\delta Q}) - \ln(1+e^{\alpha+\gamma X+\delta Q})]\right\} \\ &= \frac{1}{\beta}\ln(1+e^{\alpha+\gamma X+\delta Q}) \end{aligned} \tag{9}$$

Kriström [31] defined the spike value when A=0 as

$$Spike = \frac{1}{1 + \exp[\alpha + \gamma X + \delta Q]} \tag{10}$$

When applying the spike model, two binary variables can generally be utilised. First, M indicates whether there exists a WTP range for parties to traffic accidents (WTP>0). If a study's survey price valuation inquiries are met with a series of rejections, the WTP value may consequently be zero (A preferred fine below zero should not happen in our case). Second, *W* indicates whether price valuation is accepted when parties to traffic accidents are faced with the final bid *A*, meaning that it determines whether WTP is bigger than *A*.

$$M=\begin{cases}1, & WTP>0\\0, & WTP=0\end{cases} \tag{11}$$

$$W=\begin{cases}1, & WTP>A\\0, & others\end{cases} \tag{12}$$

Finally, the parameters of the spike model are estimated by maximum likelihood estimation, as shown below:

$$\begin{aligned}L=&\sum_{i}^{N}M_iW_i\ln[1-F_{WTP}(A)]\\&+\sum_{i}^{N}M_i(1-W_i)\ln[F_{WTP}(A)-F_{WTP}(0)]\\&+\sum_{i}^{N}(1-M_i)\ln F_{WTP}(0)\end{aligned} \tag{13}$$

DATA COLLECTION AND DATA ANALYSIS

Data Collection

The present study primarily collected survey data on WTP for the air pollution and time delays caused by traffic accidents. From 8 September to 24 September 2013, randomised sample surveys of highway travellers were conducted at rest areas along Taiwanese highways. A total of 930 surveys were issued and a total of 907 effective samples were collected. The survey was composed of three sections: (1) traffic accidents and the nature of travel, which included travel purpose, travel time, traveller identity, city or county of departure, highway junction names, *etc.*; (2) a WTP scenario, including WTP for air pollution (nitrogen dioxide and carbon dioxide) emissions and WTP for time delays; and (3) personal, household and vehicle characteristics, including gender, age, education level, occupation, personal monthly income, household monthly income, family size, physical health, *etc.*

Due to Taiwan's limited pollution source monitoring data for actual pollution and its lack of a price-setting foundation for pollution sources, this study used nitrogen dioxide and carbon dioxide as hypothetical scenarios for estimating WTP. First, by using the information provided by the Taiwan highway police agency on congestion caused by traffic accidents, based on data including processing time, the number of lanes affected and queuing distance in kilometres, we utilised the line source emission factor of TEDS7.1 (Taiwan Emission Data System, Environmental Protection Administration, Executive Yuan, Taiwan [37]) to calculate the total emissions of nitrogen dioxide and carbon dioxide, and classified them into three levels of traffic accident pollution emissions: light (<2.5 metric tons), moderate (≥2.5 metric tons and <5 metric tons) and severe (≥5 metric tons). Subsequently, we utilised the rate of Taiwan's stationary source sanction [37] to calculate the economic cost of nitrogen dioxide. In addition, the economic cost of carbon dioxide was derived from a study of "green" tax reform [38].

Finally, the starting point of A (the bid) for the three levels of pollution were provided to ask interviewees about the pollution offset price of highway traffic accidents. Three levels of the bid for light, moderate, and severe traffic accidents are NT$10,000, 35,000, and 100,000 for nitrogen dioxide emissions and NT$1000, 2500, and 8000 for carbon dioxide emissions. Please refer to Table 1 for the reasoning behind the bids.

Table 1. Three levels of bids for light, moderate, and severe traffic accidents.

Levels	NO_2 and CO_2 Emissions (Metric Tons) [a]	Economic Cost [b] of NO_2 (NT$/Metric ton)	Economic Costs of NO_2 (NT$)	Bids for NO_2 (NT$)	Economic Cost [c] of CO_2 (NT$/Metric ton)	Economic Costs of CO_2 (NT$)	Bids for CO_2 (NT$)
Light	1.0		10,000	10,000		750	1000
Moderate	3.5	10,000	35,000	35,000	750	2625	2500
Severe	10.0		100,000	100,000		7500	8000

[a]: the emission values were obtained from National Police Agency, Ministry of the Interior, Taiwan.

[b]: Environmental Protection Administration, Executive Yuan (2012) [37]. [c]: Shaw *et al.* [38].

In terms of time delays, three levels of delay scenarios were created based the average wage level of various Taiwancse industries: WTP for decreasing travel delay time caused by traffic accidents by 10, 30 and 60 min. Three levels of the bid for light (delay for 10 min), moderate (delay for 30 min), and severe (delay for one hour) traffic accidents are NT$100, 300, and 500 (The Taiwanese personal hourly salary ranged from NT$250 to 600 in 2013 [39]). It was expected that each interviewee's time delay value would be obtained from the survey, and the total time loss due to accidents can be estimated based on actual information about time delays caused by traffic accidents.

In scenarios at three levels, interviewees were provided with information such as the total pollution emissions, the affected number of lanes, the queuing distance in kilometres and the number of vehicles affected for the purpose of helping them to imagine the hypothetical situation, before asking the questions set forth below:

A. Air Pollution Scenario (Nitrogen Dioxide and Carbon Dioxide)

Taking Nitrogen Dioxide Emission of Serious Accident as an Example
Assume that the government is thinking of implementing a new policy to charge air pollution expenses caused by traffic accidents for improving the environment. Please imagine: According to Taiwan highway police agency statistics data, a serious accident may cause four lanes to be occupied, six kilometers of traffic spillback, four thousand vehicles to be affected, and nine tons of nitrogen dioxide to be emitted. In the future, if nitrogen dioxide monitors were to be set up along the highway to continually monitor nitrogen dioxide concentration levels and emissions, and if these devices were to be used to monitor severe nitrogen dioxide emissions for the purpose of sanctioning parties to traffic accidents to raise the money needed to improve air pollution caused by traffic accidents, as a party to a traffic accident, would you be willing to pay NT$100,000 to prevent the air pollution caused by that accident?

B. Time Loss Scenario
Assuming the same trip purpose and travel time, if there were to be a traffic accident on the highway, and that accident cost you 10/30/60 more minutes of travel time, would you be willing to pay $NT100/300/500 to avoid the time delays caused by the accident?

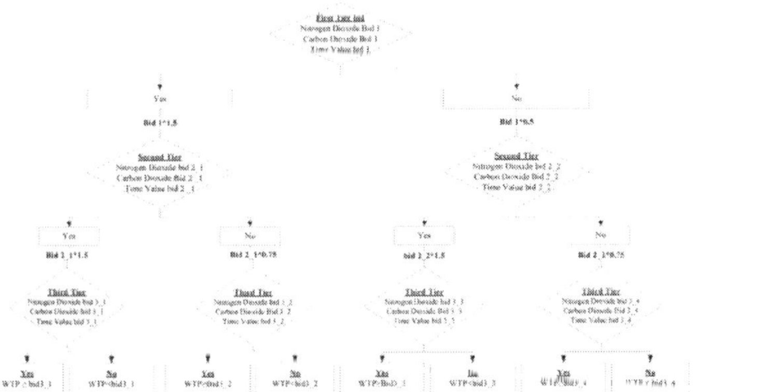

Figure 1. Triple-bound structure frame for air pollution WTP by parties to traffic accidents and time delay scenario for individuals affected by traffic accidents.

Subsequent price inquiries followed the triple-bound method (Figure 1), setting three tiers of bids, to more efficiently obtain interviewees' real WTP. In the hypothetical market scenario section, as the survey follows skip logic with three-tier questions, there were eight possible answers: (1) Yes-Yes-Yes; (2) Yes-Yes-No; (3) Yes-No-Yes; (4) Yes-No-No; (5) No-Yes-Yes; (6) No-Yes-No; (7) No-No-Yes; and (8) No-No-No. In the eighth answer scenario, the interviewee indicates no to all three questions, meaning that he/she is not willing to pay the suggested bid on the third tier, and so the next step is to ask if the interviewee's real WTP is $NT0.

Data Analysis

Socio-Economic Characteristics

Table 2 shows the interviewees' socioeconomic characteristics. The majority of the interviewees in all three levels were male, accounting for 59.5%, 52.5% and 54.6%, respectively, whereas females accounted for fewer than 50% of the interviewees. The majority of the interviewees were 31 to 40 years of age, which accounted for 32.7%, 35.2% and 33.1% of the interviewees; the next-most-represented age groups were 21 to 31 and 41 to 50 years of age. With respect to educational levels, the highest percentage of interviewees had college or junior college degrees—58.9%, 59.0% and 58.6% in each respective category—followed by interviewees holding high school diplomas or master's degrees. Occupation-wise, the service industry was the most represented among the interviewees, comprising 28.9%, 34.0% and 31.9% of the groups, followed by the commercial and industrial sectors. With respect to average personal monthly income, a plurality of interviewees earned $NT25,001–40,000—29.1%, 37.0% and 30.5%—followed by interviewees with incomes of less than $NT25,000 and $NT 40,001 to 60,000. With respect to average monthly household income, families with $NT40,001 to 60,000 represented the majority of interviewees—23.2%, 21.1% and 23.8% of each respective level—but families with incomes of $NT25,001 to 40,000, $NT60,00 to 80,000 or $NT100,001 to 150,000 were also represented in significant number. With respect to family size (including the interviewees), most of the families represented had either two or four family members, represented at 28.7% and 29.2% in the light scenario, 35.6% and 31.1% in the moderate scenario and 37.1% and 30.5% in the severe scenario.

Table 2. Socioeconomic characteristics analysis.

Item		Light		Moderate		Severe	
		Sample Number	**(%)**	**Sample Number**	**(%)**	Sample Number	**(%)**
Gender	Male	180	-59.6	159	-52.5	165	-54.6
	Female	122	-40.4	144	-47.5	137	-45.4
Age	Under 20	14	-4.7	16	-5.3	17	-5.6
	21–30	76	-25.2	79	-26.1	87	-28.8
	31–40	99	-32.7	107	-35.2	100	-33.1
	41–50	61	-20.2	65	-21.5	54	-17.9
	51–60	40	-13.2	28	-9.2	28	-9.3
	61 and above	12	-4	8	-2.7	16	-5.3
Education Level	Elementary school, Junior high school	16	-5.3	18	-5.9	18	-5.9
	(Vocational) high school	66	-21.9	89	-29.4	73	-24.2
	College/Junior college	178	-58.9	179	-59	177	-58.6
	Master's/Doctoral degree	42	-13.9	17	-5.7	32	-10.6
Occupation	Unemployed/Job-searching/Retired	15	-5	10	-3.3	24	-7.9
	Homemaker	27	-8.9	23	-7.6	30	-9.9
	Military, public or teaching	33	-10.9	21	-6.9	20	-6.6
	Agricultural, forestry, fishery or animal husbandry	3	-1	3	-1	5	-1.7
	Industry	43	-14.2	49	-16.2	52	-17.2
	Commerce	73	-24.2	63	-20.8	45	-14.9
	Service	87	-28.9	103	-34	96	-31.9
	Students	17	-5.6	20	-6.6	19	-6.3
	Others	4	-1.3	11	-3.6	11	-3.6
Average Personal Monthly Income	25,000 and below	78	-25.8	74	-26.4	88	-29.2
	25,001–40,000	88	-29.1	112	-37	92	-30.5
	40,001–60,000	77	-25.5	65	-21.5	73	-24.2

	60,001--80,000	29	-9.6	34	-11.2	24	-7.9
	80,001–100,000	13	-4.3	4	-1.3	7	-2.2
	100,001 and above	12	-3.9	8	-2.6	18	-6
Average Household Monthly Income	25,000 and below	25	-8.3	24	-8	26	-8.6
	25,001–40,000	44	-14.6	62	-20.5	46	-15.2
	40,001--60,000	70	-23.2	64	-21.1	72	-23.8
	60,001–80,000	46	-15.2	56	-18.5	45	-14.9
	80,001–100,000	34	-11.3	24	-7.9	31	-10.3
	100,001–150,000	40	-13.2	35	-11.6	42	-13.9
	150,001–200,000	16	-5.3	15	-5	20	-6.6
	200,001 and above	26	-8.6	15	-5	19	-6.3
Family Size	1 person	47	-15.6	31	-10.2	39	-12.9
	2 people	87	-28.7	108	-35.6	112	-37.1
	3 people	80	-26.5	70	-23.1	59	-19.5
	4 people and above	88	-29.2	94	-31.1	92	-30.5
Total Sample Number		302		303		302	

Note: There are two missing values (0.7%) for educational level in the severe scenario. There are five and six missing values in the light and moderate scenarios, respectively, for Personal Monthly Income; there are one, eight and one missing values in the light, moderate and severe scenarios, respectively, for Average Household Income.

Table 3 shows the interviewees' travel characteristics. Their trip purposes were mostly tourism and leisure, with shares of 49.7%, 60.4% and 55.0%, respectively, followed by visiting families and friends and business trips. The majority of highway travel time was between 120 and 180 min, with shares of 26.8%, 25.4% and 29.5%; the remaining travel times are equally distributed. Travelling time on local roads was mostly less than 30 min, with shares of 36.5%, 47.8% and 35.8%, respectively; travel times of 120 min and above were the least common. Upwards of 80% of respondents' travel was not time-sensitive; the likely reason for this result is that the study was conducted at highway rest areas, and most individuals who stopped at those rest areas were not pressed for time. With respect to traveller identity, most of the interviewees were drivers—62.6%, 54.5% and 57.6%, respectively. In addition, based on the location of the highway interchange that interviewees chose to enter/exit, we were able to further estimate highway travel distance. The distance is divided into three categories: short distances of fewer than 100 km; medium distances of between 100 and 200 km; and long distances of 300 km and above. Long distances comprised the majority of trips, with percentages of 40.1%, 41.2% and 37.1%, respectively.

Table 3. Travel characteristics analysis.

Item		Light		Moderate		Severe	
		Sample Number	**(%)**	**Sample Number**	**(%)**	**Sample Number**	**(%)**
Trip Purpose	Work commute	30	-9.9	24	-7.8	19	-6.3
	School commute	11	-3.6	8	-2.6	7	-2.3
	Business	52	-17.2	35	-11.6	44	-14.6
	Tourism/Leisure	150	-49.7	183	-60.4	166	-55
	Visiting familiesand friends	51	-17	45	-14.9	58	-19.2
	Other	8	-2.6	8	-2.7	8	-2.6
Highway Travel Time	<60 min	55	-18.2	49	-16.1	42	-13.9
	>=60 min, <120 min	45	-14.9	59	-19.5	48	-15.9
	>=120 min, <180 min	81	-26.8	77	-25.4	89	-29.5
	>=180 min, <240 min	60	-19.9	55	-18.2	61	-20.2
	>=240 min	61	-20.2	63	-20.8	62	-20.5
Local Road Travel Time	<30 min	110	-36.5	145	-47.8	108	-35.8
	>=30 min, <60 min	94	-31.1	97	-32	104	-34.4
	>=60 min, <120 min	71	-23.5	39	-12.9	50	-16.6
	>=120 min	27	-8.9	22	-7.3	40	-13.2
Traveller Identity	Driver	189	-62.6	165	-54.5	174	-57.6
	Passenger	113	-37.4	138	-45.5	128	-42.4
Highway Travel Distance	Short distance (<100 km)	91	-30.1	85	-28.1	89	-29.5
	Medium distance (>=100, <300 km)	90	-29.8	93	-30.7	101	-33.4
	Long distance (>=300 km)	121	-40.1	125	-41.2	112	-37.1
Total Sample Number		302		303		302	

Table 4 shows the interviewees' physical health and participation in environmental activities. The survey listed eight popular environmental activities in Taiwan and asked interviewees about their willingness to participate in those activities. For example, bicycle parade, environmental protection parade, environmental protection lecture, beach or mountain cleaning activities, tree planting activities, trash off the ground, environmental protection competitions and Earth Day. From the table, we can see that in the three levels, 24.2%, 25.3% and 19.5% of the interviewees did not want to participate in environmental activities; 24.5%, 21.5% and 27.2% of the interviewees indicated a desire to participate in two environmental activities; and numerous interviewees indicated a desire to participate in either one or three environmental activities. Subsequently, interviewees were asked whether they had illnesses related air pollution such as eye irritation, respiratory diseases or cardiopulmonary diseases. From the table, we can see that interviewees were most likely to report respiratory diseases, with shares of 10.9%, 6.9% and 9.6%, respectively, with the next-largest group of interviewees reporting eye irritation, with shares of 4.0%, 4.3% and 3.3%.

Table 4. Analysis of physical health and participation in environmental activities.

Item		Light		Moderate		Severe	
		Sample Number	(%)	Sample Number	(%)	Sample Number	(%)
Eye Irritation	No	290	-96	290	-95.7	292	-96.7
	Yes	12	-4	13	-4.3	10	-3.3
Respiratory Diseases	No	269	-89.1	282	-93.1	273	-90.4
	Yes	33	-10.9	21	-6.9	29	-9.6
Cardiopulmonary Diseases	No	297	-98.3	295	-97.4	301	-99.7
	Yes	5	-1.7	8	-2.6	1	-0.3
	No	73	-24.2	77	-25.3	59	-19.5
Acceptable Number of Environmental Activities	1	59	-19.5	57	-18.8	56	-18.5
	2	74	-24.5	65	-21.5	82	-27.2
	3	49	-16.2	50	-16.5	60	-19.9
	4	22	-7.3	36	-11.9	27	-8.9
	5	15	-5	10	-3.3	9	-3
	6	3	-1	4	-1.3	3	-1
	7	3	-1	2	-0.7	1	-0.3
	8	4	-1.3	2	-0.7	5	-1.7
Total Sample Number		302		303		302	

Analysis of the Interviewees' WTP

Table 5 shows WTP for nitrogen dioxide. The survey's starting points were $NT10,000, $NT35,000 and $NT100,000 for light, moderate and severe levels, respectively. The majority of interviewees reporting a WTP of NT$0 represent 34.1%, 37.3% and 34.1% at each level. The reason for this may be due to Taiwan's failure to enforce such measures; thus, most interviewees were not willing to pay the amounts listed in the survey.

Table 5. Analysis of WTP for nitrogen dioxide.

Light			Moderate			Severe		
Price	Sample Number	(%)	WTP	Sample Number	(%)	WTP	Sample Number	(%)
WTP ≥ 22,500	10	(3.3)	WTP ≥ 80,000	4	(1.3)	WTP ≥ 250,000	7	(2.3)
WTP < 22,500	33	(10.9)	WTP < 80,000	11	(3.6)	WTP < 250,000	19	(6.3)
WTP ≥ 12,000	12	(4.0)	WTP ≥ 40,000	18	(5.9)	WTP ≥ 120,000	16	(5.3)
WTP < 12,000	35	(11.6)	WTP < 40,000	35	(11.6)	WTP < 120,000	23	(7.6)
WTP ≥ 7500	9	(3.0)	WTP ≥ 30,000	10	(3.3)	WTP ≥ 70,000	10	(3.3)
WTP < 7500	35	(11.6)	WTP < 30,000	12	(4.0)	WTP < 70,000	34	(11.3)
WTP ≥ 3500	34	(11.3)	WTP ≥ 10,000	61	(20.1)	WTP ≥ 35,000	45	(14.9)
WTP < 3500	31	(10.2)	WTP < 10,000	39	(12.9)	WTP < 35,000	45	(14.9)
WTP = 0	103	(34.1)	WTP = 0	113	(37.3)	WTP = 0	103	(34.1)
Sample Number	302		Sample Number	303		Sample Number	302	

Table 6 analyses WTP for carbon dioxide. The starting points of the survey were $NT1000, $NT2500 and $NT8000 for the light, moderate and severe levels, respectively. The majority WTP was NT$0, which were 31.1%, 34.3% and 35.1% at each level.

Table 6. Analysis of WTP for carbon dioxide.

Light			Moderate			Severe		
Price	Sample Number	(%)	WTP	Sample Number	(%)	WTP	Sample Number	(%)
WTP ≥ 2500	46	(15.3)	WTP ≥ 6000	23	(7.6)	WTP ≥ 18,000	21	(7.0)
WTP < 2500	42	(13.7)	WTP < 6000	29	(9.6)	WTP < 18,000	31	(10.3)
WTP ≥ 1200	18	(6.0)	WTP ≥ 3000	24	(7.9)	WTP ≥ 9,000	26	(8.6)
WTP < 1200	50	(16.6)	WTP < 3000	41	(13.5)	WTP < 9,000	34	(11.3)
WTP ≥ 750	7	(2.3)	WTP ≥ 2000	7	(2.3)	WTP ≥ 6,000	11	(3.6)
WTP < 750	19	(6.3)	WTP < 2000	13	(4.3)	WTP < 6,000	22	(7.3)
WTP ≥ 350	21	(7.0)	WTP ≥ 1000	47	(15.5)	WTP ≥ 3,000	27	(8.9)
WTP < 350	5	(1.7)	WTP < 1000	15	(5.0)	WTP < 3,000	24	(7.9)
WTP = 0	94	(31.1)	WTP = 0	104	(34.3)	WTP = 0	106	(35.1)
Sample Number	302		Sample Number	303		Sample Number	302	

Table 7 shows WTP for time delays. The survey's starting point for time delays caused by light, moderate and severe traffic accidents were $NT100, $NT300 and $NT500, respectively. The majority WTP was also $NT0, with shares of 31.5%, 32.7% and 26.2%.

Table 7. Analysis of WTP for time delays.

Light			Moderate			Severe		
Price	Sample Number	(%)	WTP	Sample Number	(%)	WTP	Sample Number	(%)
WTP ≥ 250	52	(17.2)	WTP ≥ 700	42	(13.9)	WTP ≥ 1200	31	(13.3)
WTP < 250	74	(24.5)	WTP < 700	44	(14.5)	WTP < 1200	55	(18.2)
WTP ≥ 125	17	(5.6)	WTP ≥ 350	34	(11.2)	WTP ≥ 600	29	(9.6)
WTP < 125	39	(12.9)	WTP < 350	35	(11.6)	WTP < 600	59	(19.5)
WTP ≥ 75	3	(1.0)	WTP ≥ 250	10	(3.3)	WTP ≥ 350	6	(2.0)
WTP < 75	5	(1.6)	WTP<250	14	(4.6)	WTP < 350	17	(5.6)
WTP ≥ 35	12	(4.0)	WTP ≥ 100	14	(4.6)	WTP ≥ 150	16	(5.3)
WTP < 35	5	(1.7)	WTP < 100	11	(3.6)	WTP < 150	10	(3.3)
WTP = 0	95	(31.5)	WTP = 0	99	(32.7)	WTP = 0	79	(26.2)
Sample Number	302		Sample Number	303		Sample Number	302	

ESTIMATION RESULTS

Variable Descriptions

Different variables were evaluated in the model; only those variables with significant estimation results are listed and discussed herein. From Table 8, we see that the dependent variables are WTP for nitrogen dioxide, carbon dioxide and time delays, and that the WTP range is as follows. $NT0–250,000 for nitrogen dioxide, $NT0–100,000 for carbon dioxide and $NT0–12,000 for time delays. The independent variables in this model estimation are dummy variables. In the hypothetical scenario for nitrogen dioxide, based on *a priori* knowledge, variables with a significant positive effect in the estimation model are as follows: "age 41 years and above"; "Education level of college or above"; "Willing to participate in two or more environmental activities"; "Eye irritation and respiratory diseases"; "Regular employee and average personal monthly income of NT$25,000 or more"; "Married and has a family size of two or fewer"; and "Willing to participate in environmental activities and average monthly household income is NT$60,000 or more".

In the case of carbon dioxide, variables that have a significant positive effect in the estimation model include "Age between 31and 60"; "Education level of college or above"; "Personal monthly income of NT$40,000 and above"; "Eye irritation and respiratory diseases"; "Willing to participate in one or two environmental activities"; "Age 51 years or above and willing to participate in environmental activities ", and "Education level of college or above and average monthly household income is NT$60,000 or above".

For time delays, variables that have a significant positive effect in this estimation model include "managerial rank"; "work/school commute"; "travelling as the driver"; "two hours or more highway travel time"; "long travel distance (≥100 km) on this trip"; "short travel distance (<100 km) on this trip". The model estimation was conducted based on the variables in the table.

Results

This section includes the WTP estimation model results for nitrogen dioxide, carbon dioxide and time delays, which are discussed separately. From the model estimation, we can see that the spike values shown in the estimation results are the WTP of the sample data. The WTP estimation results for nitrogen dioxide are shown in Table 9. The estimation result shows that WTP per metric ton is NT$11,502 for the light level, NT$9004 for the moderate level and NT$8862 for the severe level. Based on the unit cost of varied serious level, while the air pollution is more serious and the total value of WTP of interviewees increases, the relative average cost (NT$/metric ton) decreases.

Table 8. Variable definitions.

Dependent Variable

Variable	Range
1. Nitrogen dioxide scenarios	Light: 0–100,000
	Moderate: 0–100,000
	Severe: 0–250,000
2. Carbon dioxide scenarios	Light: 0–50,000
	Moderate: 0–100,000
	Severe: 0–100,000
3. Time delay scenarios	Light: 0–5,000
	Moderate: 0–10,000
	Severe: 0–12,000

Independent Variable

Variable	Range	Expectation
Age 41 years and above	1, otherwise 0	1
Age between 31and 60	1, otherwise 0	2
Education level of college or above	1, otherwise 0	+[1, 2]
Personal monthly income of NT$40,000 and above	1, otherwise 0	2
Willing to participate in two or more environmental activities	1, otherwise 0	1
Eye irritation and respiratory diseases	1, otherwise 0	+[1,2]
Willing to participate in one or two environmental activities	1, otherwise 0	2
Managerial rank	1, otherwise 0	3
Work/school commute	1, otherwise 0	3
Travelling as the driver	1, otherwise 0	3
Travel time of two hours or more for this trip	1, otherwise 0	3
Long travel distance (≥100 km) for this travel	1, otherwise 0	3
Short travel distance (<100 km) for this travel	1, otherwise 0	3

Interactive Variable

Regular employee and average personal monthly income of NT$25,000 or more	1, otherwise 0	1
Married and has a family size of two or fewer	1, otherwise 0	1

Willing to participate in environmental activities and average monthly household income is NT$60,000 or more	1, otherwise 0	1
Age 51 years or above and willing to participate in environmental activities	1, otherwise 0	2
Education level of college or above and average monthly household income is NT$60,000 or above	1, otherwise 0	2

Notes: 1 indicates nitrogen dioxide scenarios; 2 indicates carbon dioxide scenarios; 3 indicates time delay scenarios.

In the estimation model, the negative coefficient of bid shows that for those interviewees who are willing to participate in the nitrogen dioxide pollution offset, the higher the bid, the more interviewees are reluctant to pay. This result is consistent with the expectations of the present study. From the table, we see that the variables that have significant positive effects include "age 40 years or above", "college/junior college degree or above", "eye irritation or respiratory diseases", and "willing to participate in two or more environmental activities". Interviewees who are 40 years of age or older and who have a college/junior college degree or above have greater WTP. This is likely because these groups of travellers have better comprehension and judgment and therefore are more willing to pay for pollution emissions caused by traffic accidents for which they are responsible.

Table 9. Model estimation result of WTP for nitrogen dioxide.

Variable	Light		Moderate		Severe	
	Coefficient	*t*-value	Coefficient	*t*-value	Coefficient	*t*-value
Constant	0.03	(−0.20)	0.12	-0.83	−0.14	(−0.94)
Bid	−0.83	(−6.46 **)	−0.36	(−7.31 **)	−0.14	(−9.51 **)
Age 41 years or above	0.6	(2.90 **)	0.37	(1.92 *)	—	—
College/junior college degree or above	0.51	(2.65 **)	—	—	0.91	(5.14 **)
Eye irritation or respiratory diseases	0.82	(1.99 **)	—	—	2	(2.05 **)
Willing to participate in two or more environmental activities	0.67	(2.19 **)	—	—	—	—

Married with family size of two or fewer	–	–	0.79	(1.92 *)	–	–
Willing to participate in environmental activities and average monthly household income is NT$60,000 or above	–	–	0.73	(1.71 *)	0.81	(1.91 *)
Regular employee and average personal monthly income of NT$25,000 or more	–	–	0.36	(2.27 **)	–	–
WTP (NT$/metric ton)	11,502	(5.57 **)	9,004	(6.29 **)	8,862	(7.40 **)
Spike value	0.38	(9.29 **)	0.38	(10.05 **)	0.32	(9.93 **)
Wald statistic (p-value)	9.29		285.33		405.96	
Log likelihood function	306.60		−324.82		−318.53	
Sample number	302		303		302	

Notes: * indicates significant level of 0.1; ** indicates significant level of 0.05.

Additionally, it is reasonable to assume that interviewees who are willing to participate in two or more environmental activities are more environmentally conscious and thus are more willing to pay a penalty for nitrogen dioxide emissions caused by traffic accidents. Furthermore, when these type of interviewees also have illnesses such as eye irritation or respiratory diseases, they are more likely to be sensitive to air pollution conditions than are interviewees who do not have such symptoms, in turn increasing the first group's WTP. It is evident that the coefficient value 0.82 has the most significant effect in the model. The estimation result of this model is in accordance with the results shown in the study of Wang and Zhang [13], which indicate that household income/expense, respiratory diseases, education level, and opinions on air pollution and health-related issues are the significant variables that influence air quality improvement.

In moderate and severe levels, more interactive variables are included and are significant statistically. It shall be noted that compensation amounts rise as the pollution level rises. Therefore, interviewees with lower socioeconomic backgrounds may not be able to pay fees; interviewees must be environmentally conscious and also have a high income to be able to pay higher compensation. For example, the variables "Married and has a family size of two or fewer", "Willing to participate in environmental activities and average monthly household income is NT$60,000 or above" and "Regular employee and has an average monthly income of NT$25,000 or above" have positive and significant effects at the moderate level of pollution caused by traffic accidents. We also assume that interviewees who are willing to participate in environmental activities feel obligated to pay higher amounts to compensate for nitrogen dioxide pollution for which he/she is responsible. However, when the pollution level is moderate or severe, we do not consider willingness to participate in environmental activities, interviewees must have an average household monthly income of NT$60,000 to demonstrate a higher WTP to compensate for nitrogen dioxide emissions caused by traffic accidents, meaning that only those interviewees whose families have superior economic well-being at a certain level can afford and are willing to pay compensation. In order to afford the pollution compensation fee, interviewees may have to be married and have fewer than two family members (without the economic burden of child support). Subsequently, compared to part-time workers, full-time employees have more secure employment, and only with an average monthly income of more than NT$25,000 can interviewees afford the fee, and thus have a higher WTP.

Table 10 shows the analysis of the WTP estimation results for carbon dioxide. The average WTP is NT$2693 in the light level, NT$1120 in the moderate level, and NT$1070 in the severe level. From the table, we can see that the significant positive effects in the light level belong to variables such as "personal monthly income of NT$40,000 or above", "college/junior college degree or above", "Age between 31 and 60 years" and "Willing to participate in two or two environmental activities". This result is most likely because an average personal monthly income of NT$40,000 is almost equal the national average personal monthly income of approximately NT$44,798 (Accounting and Statistics, Executive Yuan, R.O.C., 2013) [10], and therefore, interviewees with this income level can afford the extra expense and are more willing to pay carbon dioxide compensation fees triggered by traffic accidents. The other three variables are similar to the explanations shown in the nitrogen dioxide estimation results: it is likely because these interviewees have higher comprehension and judgment abilities and pay more attention to the greenhouse effect caused by carbon dioxide emissions; in other words, they are environmentally conscious and therefore are more willing to pay for carbon dioxide caused by traffic accidents.

In terms of the moderate level, we can see that interviewees with eye irritation and respiratory diseases, "Age between 31 and 60 years" and "Willing to participate in two or two environmental activities" are more willing to pay compensation fees for carbon dioxide. The reasons are similar to the explanations mentioned earlier. Because a severe pollution level requires a higher compensation fee, only the variables of "eye irritation or respiratory diseases" and "Age 51 years or above and willing to participate in environmental activities" show a significant positive effect on WTP. Moreover, from the coefficient value of 0.79, we can see that this has the biggest impact on the model. Therefore, this group of interviewees should be the target audience for any future promotion of this policy. Finally, interviewees must be well-educated and also have a high income to be able to pay higher compensation. For example, the variable "Education level of college or above and average monthly household income is NT$60,000 or above" has positive and significant effects at the severe level of pollution caused by traffic accidents.

Table 10. Model estimation result of WTP for carbon dioxide.

Variable	Light		Moderate		Severe	
	Coefficient	***t*-value**	**Coefficient**	***t*-value**	**Coefficient**	***t*-value**
Constant	−0.08	(−0.40)	0.18	-1.18	0.01	-0.09
Bid	−0.53	(−4.85 **)	−0.29	(−7.02 **)	−0.12	(−8.42 **)
Age between 31 and 60	0.35	(1.92 *)	0.37	(2.33 **)	—	—
Education level of college or above	0.41	(2.01 **)	—	—	—	—
Personal monthly income of NT$40,000 or above	0.46	(1.99 **)	—	—	—	—
Eye irritation or respiratory diseases	—	—	1.11	(1.81 *)	0.72	(2.11 **)
Willing to participate in one or two environmental activities	0.37	(1.79*)	0.34	(1.94 *)	—	—

Age 51 years or above and willing to participate in environmental activities	–	–	–	–	0.79	(1.96 **)
Education level of college or above and average monthly household income is NT$60,000 or above	–	–	–	–	0.38	(1.74 *)
WTP(NT$/metric ton)	2,693	(4.10 **)	1,120	(6.34 **)	1,070	(4.27 **)
Spike value	0.32	(5.22 **)	0.37	(7.40 **)	0.32	(4.32 **)
Wald statistic (*p*-value)	283.87		205.14		267.26	
Log likelihood function	−331.58		−331.82		−316.22	
Sample number	302		303		302	

Notes: * indicates significant level of 0.1; ** indicates significant level of 0.05.

The WTP estimation results for time delays are shown in Table 11. From the table, we can see that WTP for one hour is NT$1315 for the light level, NT$1027 for the moderate level and NT$960 for the severe level. Overall, travellers identifying as drivers has a positive effect on WTP for time delays caused by traffic accidents at three levels. This is most likely because interviewees who are drivers had stronger and direct feelings about driving time and therefore were more willing to pay. In addition, work/school commute and short travelling distance also have positive effects only on the light level. The result of short travelling distance is possibly because travel time is comparatively short in short-distance travel, making a 10-minute delay represent a large proportion of the entire trip (32.2% approximately); therefore, short-distance travellers are willing to pay to avoid time delays.

Table 11. Model estimation results of WTP for time delays.

Item	Light		Moderate		Severe	
	Coefficient	***t*-value**	**Coefficient**	***t*-value**	**Coefficient**	***t*-value**
Constant	0.1	-0.62	0.39	(2.37 **)	0.25	(1.72 *)
Bid	−5.37	(−6.95 **)	−2.11	(−6.82 **)	−1.25	(−14.24 **)
Traveller is driver	0.71	(3.61 **)	0.31	(1.73 *)	0.3	(1.65 *)
Work/school commute	0.39	(1.83 *)	–	–	–	–
Travel time of two hours or more for this trip	–	–	0.31	(1.69 *)	0.58	(2.64 **)
Medium highway travel distance for this trip (≥100 km, <300 km)	–	–	0.45	(1.96 **)	0.88	(3.71 **)
Short highway travel distance for this trip (<100 km)	0.77	(3.29 *)	–	–	–	–
Managerial rank	–	–	–	–	0.6	(1.75 *)
WTP (NT$/hour)	1315	(5.98 **)	1,027	(6.41 **)	960	(6.95 **)
Spike value	0.31	(5.98 **)	0.34	(8.02 **)	0.3	(5.94 **)
Wald statistic (*p*-value)	206.68		189.36		683.25	
Log likelihood function	−342.71		−338.97		−343.92	
Sample number	302		303		302	

Note: * indicates significant level of 0.1; ** indicates significant level of 0.05.

In the moderate and severe levels, interviewees whose travel time was two hours or more were more willing to pay for time delays caused by traffic accidents, possibly trying to avoid longer travel times. Interviewees with medium travelling distance have a positive effect on WTP in the moderate and severe levels. Travelling time is naturally longer for medium-distance travellers, and therefore a 10-min delay in travelling time has relatively little impact. Therefore, medium-distance travellers only have a stronger WTP when time delays are more serious in moderate and severe levels, *i.e.*, when time delays equal 30 or 60 min. Finally, in the severe level, interviewees at the managerial level have a stronger WTP to avoid time delays. This is likely because the value of these interviewees' time is higher due to their professional rank; therefore, when faced with a long time delay, they are more willing to pay a higher amount to prevent the delay from happening.

Compared the WTP for time loss to other studies, the results show that Taiwanese value of time is close to the United States (as shown in Table 12).

Table 12. Comparisons of value of time.

Sources	Country	Value of Time (US$/hour)
Our study	Taiwan	32–43
Ghosh [40]	U.S.A.	20–57
Brownstone [41]	U.S.A.	23–43
Steimetz and Brownstone[42]	U.S.A.	46

CONCLUSIONS AND RECOMMENDATIONS

This study adopted the TBDC method to estimate WTP. Interviewees' WTPs were obtained using three tiers of price inquiry. The spike model was used to estimate the WTP for air pollution due to traffic accidents caused by the parties, along with the affected individuals' WTP for time delays due to traffic accidents caused by them. This section proposes conclusions and recommendations based on this study's research findings.

From the WTP estimation model of nitrogen dioxide and carbon dioxide, we can see that age, income, education level, willingness to participate in environmental activities and cross-variables such as average monthly income combined with willingness to participate in environmental activities affect the results of the model. In both models, as the severity of pollution and the compensation amount both rise, the interviewees' WTP is influenced by more than a single socioeconomic characteristic; multiple characteristics such as high income or environmental consciousness are required for a positive effect on WTP. In the time delays model, variables such as occupational rank, trip purpose, traveller identity, travel time and travel distance have positive effects on WTP. Because the severity of traffic accidents differs, the level of air pollution and time loss differ significantly. The interviewees' WTP also differs depending on the level of severity. In the case of nitrogen dioxide, the average WTP is NT$11,502 for the light level, NT$9004 for the moderate level and NT$8862 for the severe level. For carbon dioxide, the average WTP is NT$2693 for the light level, NT$1120 for the moderate level and NT$1070 for the severe level. With respect to time loss, the average WTP for an hour is NT$1315 for light congestion, $1027 for moderate congestion and $960 for severe congestion.

However, in Taiwan there is no standard method to measure air pollution caused by traffic accidents, and the severity (light, moderate or severe) of pollution was not included in the traffic accident database. With respect to time delays caused by traffic accidents, only the overview of traffic accidents (including estimated time delay and distance) gathered by the Taiwanese Police Radio Station was collected, and there were no complete traffic accident investigation records that could be utilised. Therefore, we suggest that in the future, traffic accident data should include factors such as the air pollution, processing time and amount of traffic congestion caused by traffic accidents. Only through determining the different types of pollution levels and time delays caused by traffic accidents can different compensation levels be calculated to provide a reference point for the external cost of traffic accidents.

We believe that individuals who cause traffic accidents should compensate for the impact of air pollution and time delays caused by those accidents, a position that is consistent with the polluter-pays principle—utilising the income of the polluter to compensate for air pollution and time loss. Although this measure has not been implemented, this study obtains WTP for air pollution by individuals who cause traffic accidents, which in turn provides a reference for relevant agencies to establish an applicable sanction standard. However, this study is the first attempt to study an air pollution and time loss compensation amount targeted at the parties or individuals in traffic accidents; therefore, the research area is confined to highway users. We recommend that future studies be expanded to local roads for the purpose of obtaining more complete research data.

ACKNOWLEDGMENTS

This study was supported by the Ministry of Science and Technology (Project No. MOST 103-2811-E-260-001).

AUTHOR CONTRIBUTIONS

Rong-Chang Jou designed the study and writing of the paper. Tzu-Ying Chen analyzed the data, conducted the research, writing of the paper and responded to the comments raised by the reviewers. Both the co-authors drafted and revised the manuscript together. Both authors read and approved the final manuscript.

REFERENCES AND NOTES

1. Jou, R.C. The amount of consolation compensation in road traffic accidents. *Accid. Anal. Prev.* **2014**, *67*, 21–29.
2. Elvik, R. The external costs of traffic injury: Definition, estimation, and possibilities for internalization. *Accid. Anal. Prev.* **1994**, *26*, 719–732.
3. Forkenbrock, D.J. External costs of intercity truck freight transportation. *Transp. Res. A* **1999**, *33*, 505–526.
4. Zegras, C. The costs of transportation in Santiago de Chile: Analysis and policy implications. *Transp. Policy* **1998**, *5*, 9–21.
5. Shafie-Pour, M.; Ardestani, M. Environmental damage costs in Iran by the energy sector. *Energy Policy* **2007**, *35*, 4413–4423.
6. Wang, X.J.; Zhang, W.; Li, Y.; Yang, K.Z.; Bai, M. Air quality improvement estimation and assessment using contingent valuation method, a case study in Beijing. *Environ. Monit. Assess.* **2006**, *120*, 153–168.
7. Cai, C.G.; Zheng, X.Y. Application of contingent valuation method in valuing health gains from air quality improvement. *Res. Environ. Sci.* **2007**, *20*, 150–154.
8. Carlsson, F.; Kataria, M.; Krupnick, A.; Lampi, E.; Löfgren, Å.; Qin, P.; Chung, S.; Sterner, T. Paying for mitigation: A multiple country study. *Land Econom.* **2012**, *88*, 326–340.
9. Lera-López, F.; Faulin, J.; Sánchez, M. Determinants of willingness-to-pay for reducing the environmental impacts of road transportation. *Transp Res. Part D* **2012**, *17*, 215–220.
10. Kotchena, M.J.; Boylec, K.J.; Leiserowitz, A.A. Willingness-to-pay and policy-instrument choice for climate-change policy in the United States. *Energy Policy* **2013**, *55*, 617–625.
11. Peng, X.Z.; Tian, W.H. Study on willingness to pay for air pollution on economic loss in Shanghai. *World Econom. Forum* **2003**, *3*, 32–44.

12. Wang, H.; Mullahy, J. Willingness to pay for reducing fatal risk by improving air quality: A contingent valuation study in Chongqing, China. *Sci. Total Environ.* **2006**, *367*, 50–57.
13. Wang, Y.; Zhang, Y.S. Air quality assessment by contingent valuation in Ji'nan, China. *J. Environ. Manag.* **2009**, *90*, 1022–1029.
14. Bhat, C.R.; Sardesai, R. The impact of stop-making and travel time reliability. *Transp. Res. Part A* **2006**, *40*, 709–730.
15. Senna, L.A.D.S. The influcnce of travel-time variability on the value of time. *Transportation* **1994**,*21*, 203–228.
16. Hcnsher, D.A. Measurement of the valuation of travel-time savings. *J. Transp. Econom. Policy* **2001**, *35*, 71–98.
17. Hensher, D.A. The valuation of commuter travel-time savings for car drivers: Evaluating alternative model specifications. *Transportation* **2001**, *28*, 101–118.
18. Li, Z.; Hensher, D.A.; Rose, J.M. Willingness to pay for travel time reliability in passenger transport: A review and some new empirical evidence. *Transp. Res. Part E* **2010**, *46*, 384–403.
19. Ciriacy-Wantrup, S.V. Capital Returns from Soil-Conservation Practices. *J. Farm Econom.* **1947**,*29*, 1181–1996.
20. Thunberg, E.; Shabman, L. Determinants of Landowners' Willingness-to-Pay for Flood Hazard Reduction. *Water Resourc. Bull.* **1991**, *27*, 657–665.
21. Davies, J.C.; Mazurek, J. *Pollution Control in the United States: Evaluating the System (Resources for the Future)*; Routledge: London, UK, 1998.
22. Tobin, J. Liquidity Preferences as Behavior Towards Risk. *Rev. Econ. Stud.* **1958**, *25*, 65–86.
23. Wodjao, T.B. A double-hurdle model of computer and internet use in American households. Available online: http://www.atususers.umd.edu/wip2/papers_i2007/Wodajo.pdf (accessed on 29 May 2015).
24. Anastasopoulos, P.C.; Tarko, A.P.; Mannering, F.L. Tobit analysis ofvehicle accident rates on interstate highways. *Accid. Anal. Prev.* **2008**, *40*, 768–775.
25. Botzena, W.J.W.; van den Bergh, J.C.J.M. Risk attitudes to low-probability climate change risks: WTP for flood insurance. *J. Econom. Behav. Organ.* **2012**, *82*, 151–166.
26. Xu, X.; Kouhpanejade, A.; Saric, Z. Analysis of influencing factors identification of crash rates using tobit model with endogenous variable. *Promet Traffic Transp.* **2013**, *25*, 217–224.
27. Halstead, J.M.; Luloff, A.E.; Stevens, T.H. Protest bidders in contingent valuation. *Northeast J. Agric. Resour. Econom.* **1992**, *21*, 160–169.

28. Whitehead, J.C., Groothuis, P.A., Blomquist, G.C. Testing for non-response and sample selection bias in contingent valuation: Analysis of a combination phone/mail survey. *Econom. Lett.* **1993**,*41*, 215–220.
29. Freeman, A.M. On assessing the state of the art of the contingent valuation method of valuing environmental changes. In *Valuing Environmental Goods: An Assessment of the Contingent Valuation Method*; Cummings, R.G., Brookshire, D.S., Schulze, W.D., Eds.; Rowman & Allanheld: Totowa, NJ, USA, 1986.
30. Reichl, J.; Frühwirth-Schnatter, S. A censored random coefficients model for the detection of zero willingness to pay. *Quant. Market. Econom.* **2012**, *10*, 259–281.
31. Kriström, B. Spike models in contingent valuation. *Am. J. Agric. Econom.* **1997**, *79*, 1013–1023.
32. Lim, H.-J.; Yoo, S.-H. Train travel passengers' willingness to pay to offset their CO_2 emissions in Korea. *Renew. Sustain. Energy Rev.* **2014**, *32*, 526–531.
33. Yoo, S.H.; Shin, C.O.; Kwak, S.J. Inconvenience cost of spam mail: A contingent valuation study.*Appl. Econom. Lett.* **2006**, *13*, 933–936.
34. Bengochea-Morancho, A.; Fuertes-Eugenio, A.M.; del Saz-Salazar, S. A comparison of empirical models used to infer the willingness to pay in contingent valuation. *Empir. Econom.* **2005**, *30*, 235–244.
35. Jou, R.C.; Wu, Y.C.; Chen, K.H. Analysis of the environmental benefit of a motorcycle idling stop policy at urban intersection. *Transportation* **2011**, *38*, 1017–1033.
36. Hanemann, W.M. Welfare evaluations in contingent valuation experiments with discrete responses. *Am. J. Agric. Econom.* **1984**, *66*, 332–341.
37. Air Pollution Prevention Fee Program in Taiwan. Available online: http://ivy5.epa.gov.tw/epalaw/docfile/040260.doc (accessed on 3 June 2015).
38. Green Tax Reform. Taxation Administration, Ministry of Finance, R.O.C., 2012. Available online:http://www.cier.edu.tw/ct.asp?xItem=11407&ctNode=11&mp=1 (accessed on 29 May 2015).
39. Directorate-General of Budget, Accounting and Statistics, Executive Yuan, R.O.C. 2013. Statistical Tables of Establishment Survey.
40. Ghosh, A. Valuing Time and Reliability: Commuters' Mode Choice from a Real Time Congestion Pricing Experiment. Ph.D. Thesis, Department of Economics, University of California at Irvine, Santa Monica, CA, USA, 2001.

41. Brownstone, D.; Ghosh, A.; Golob, T.F.; Kazimi, C.; van Amelsfort, D. Drivers willingness-to-pay to reduce travel time: Evidence from the San Diego I-15 Congestion Pricing Project. *Transp. Res. A* **2003**, *37*, 373–387.
42. Steimetz, S.S.C.; Brownstone, D. Estimating commuters' "Value of time" with noisy data: A multiple imputation approach. *Transp. Res. B* **2005**, *39*, 865–889.

CITATION

Rong-Chang Jou and Tzu-Ying Chen, External Costs to Parties Involved in Highway Traffic Accidents: The Perspective of Highway Users, doi:10.3390/su7067310.

CHAPTER 6

Force-Based and Displacement-Based Reliability Assessment Approaches for Highway Bridges Under Multiple Hazard Actions

Chao Huang[1,2], *Jerry J. Shen*[2], *Mi Zhou*[3,4], *George C. Lee*[1]

[1] Department of Civil, Structural and Environmental Engineering, University at Buffalo, Buffalo 14260, USA

[2] Genex Systems/TFHRC, 6300 Georgetown Pike, McLean 22101, USA

[3] National Engineering and Research Center for Highways in Mountain Area, Chongqing 400067, China

[4] Key Laboratory for Old Bridge Detection and Reinforcement Technology of Ministry of Transportation, Chang'an University, Xi'an 710064, China

ABSTRACT

The strength limit state of American Association of State Highway and Transportation Officials (AASHTO) Load and Resistance Factor Design (LRFD) Bridge Design Specifications is developed based on the failure probabilities of the combination of non-extreme loads. The proposed design limit state equation (DLSE) has been fully calibrated for dead load and live load by using the reliability-based approach. On the other hand, most of DLSEs in other limit states, including the extreme events Ⅰ and Ⅱ, have not been developed and calibrated though taking certain probability-based concepts into account. This paper presents an assessment procedure of highway bridge reliabilities under the limit state of extreme event Ⅰ, i. e., the combination of dead load, live load and earthquake load. A force-based approach and a displacement-based approach are proposed and implemented on a set of nine simplified bridge models. Results show that the displacement-based approach comes up with more convergent and accurate reliabilities for selected models, which can be applied to other hazards.

INTRODUCTION

Designing bridges to resist extreme hazard loads has been a major concern of American Association of State Highway and Transportation Officials (AASHTO) and the bridge engineering community for decades. In recent decades, a considerable amount of efforts were devoted to earthquake and wind effects and also extended to other hazards, such as scour, storm surge, vessel and vehicular collisions, etc. (Huang et al., 2014,Kameshwar and Padgett, 2014a, Kameshwar and Padgett, 2014b, Lee et al., 2008, Lee et al., 2011, Lee et al., 2013a, Lee et al., 2013b, Lee et al., 2014, Li et al., 2012, Liang and Lee, 2012, Liang and Lee, 2013, Wang et al., 2014a, Wang et al., 2014b, Wang et al., 2014c and Zhu and Frangopol, 2013). Other hazards including fire and blasting were also taken into account in the multiple hazard framework (Fujikura and Bruneau, 2012,Petrini and Palmeri, 2012, Potra and Simiu, 2009 and Rini and Lamont, 2008). Analysis and design for these hazard loads have been developed individually because each one requires different expertises and knowledges to formulate the "demand" or "capacity" for bridge design. However, a bridge may experience these multiple types of hazards at the same time or in a certain sequence. Examples include the bridge damage during 2011 Tohoku earthquake and the tsunami in Japan (Akiyama et al., 2013), and the combination of earthquake and mudflow after 2008 Wenchuan earthquake (Cui et al., 2008). The bridge capacity needs to be carefully designed by considering its resistance to these multiple extreme hazards in a fair platform to ensure the safety of the bridge structure and the transportation system.

The establishment and implementation of AASHTO Load and Resistance Factor Design (LRFD) Bridge Design Specifications (AASHTO, 2012) were considered as a milestone in the historical development of bridge design specifications. In AASHTO LRFD, both loads and resistance were treated as random variables to take the uncertainties introduced by various external or internal factors into account. For the first time, the safety of bridge was evaluated in a quantitative approach using the bridge reliabilities or the risks against various loads and/or load combinations.

AASHTO LRFD can be categorized to four types of design limit states: strength, service, extreme events and fatigue. The calibration of AASHTO LRFD consisted of trying out various combinations of load and resistance factors on a number of bridges and their components (AASHTO, 2012). Currently the strength design limit state in the AASHTO LRFD is formulated and fully calibrated using the reliability-based approach (Kulicki, 1998, Kulicki et al., 2007 and Nowak, 1999). The extreme event design limit states, however, are constructed by combining the non-extreme load effects with the independently established extreme hazard load effects through professional judgment. Its margin of safety and adequacy has not been assessed quantitatively.

Lee et al., 2008 and Lee et al., 2011 highlighted the problems related to the standards indications in order to obtain commeasurable criteria of multiple hazards for highway bridge design. To face these issues the Multidisciplinary Center for Earthquake Engineering Research (MCEER) in 2009 started a project for developing uniform risk design criteria for highway bridges under multiple hazards.

Lee et al., 2013a and Lee et al., 2013b and Huang et al. (2014) presented several different approaches and theoretical backgrounds for the establishment of design limit state equations (DLSEs) with a comprehensive consideration of multiple hazards based on proper reliability measures. Sun et al. (2015) applied the proposed method on a typical bridge located at Southeast China to calibrate the risk of failure. A particular force-based approach was presented to combine two time-varying loads with the permanent load. In this paper, with limited data/resources, the application of this approach and a displacement-based approach are given to demonstrate the procedure for calibrating the bridge reliability under dead load, live load and earthquake. Furthermore, the results are compared to address the needs for obtaining a more accurate result from the standpoint of bridge engineering community.

GENERAL STEPS FOR EVALUATION OF BRIDGE RELIABILITY

Based on the previous research by Nowak (1999), the entire procedure to establish the current AASHTO LRFD limit states consists of six steps (Lee et al., 2013a and Lee et al., 2013b). A flow chart to demonstrate the general procedure to determine bridge reliability is given in Fig. 1.

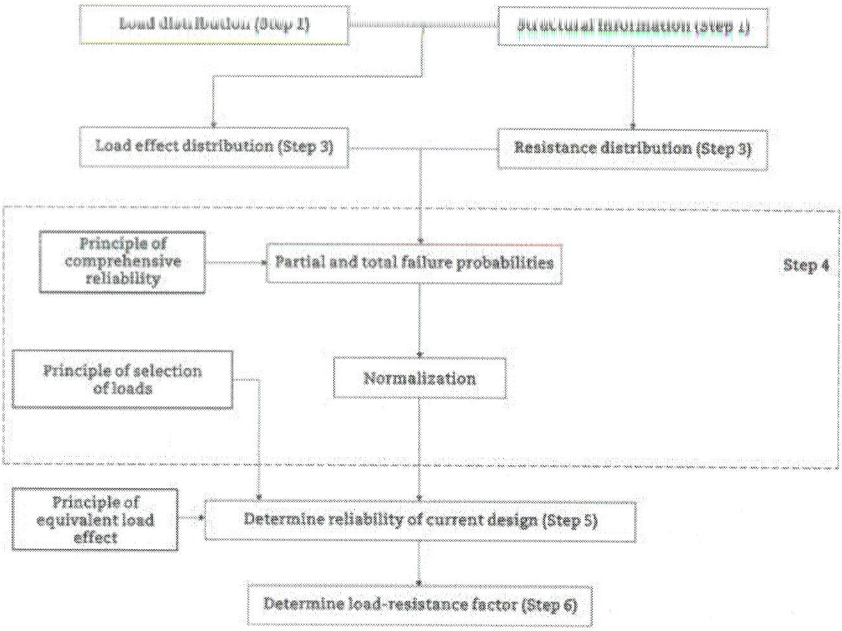

Figure 1. Flow chart for determination of load factor.

The first four steps can be used to evaluate the bridge reliability, as following:

1) Selection of representative bridges. Bridges vary with many parameters, such as structural system, material, span, width, height, skew, etc. To obtain the reasonable result, representative bridges need to be selected to conduct further analysis. Since the considered loads are dead load, live load and earthquake, and the earthquake load effect is mainly applied on the substructure of a bridge, the research focus in this study is placed on the bridge columns with the assumption that other bridge components can be simplified.
2) Establishing the statistical database for load and resistance parameters. Load distributions can be obtained from recorded data and statistical analysis, which commonly reflect the probability of occurrence and related intensity for each hazard. Many researchers proposed reasonable assumptions for simplicity. These simplified assumptions can be used for calculation before more accurate data is obtained.
3) Development of load and resistance models. For each given bridge structure, we can calculate three items: the load effect distribution, the nominal value of load effect (Q_1, Q_2, Q_3, ...) and the resistance

distribution. Note that the load effect distribution is different from the load distribution, and the failure probability can only be calculated through load effect distribution. For example, the load distribution for earthquake commonly referred to is the distribution of peak ground acceleration (PGA), but the failure probability must be calculated by using the internal force such as bending moment applied on the column.

4) Development of the reliability analysis procedure. Borges and Castanheta, 1972,Wen, 1977, Turkstra and Madsen, 1980, Ghosn et al., 2003 and Liang and Lee, 2012proposed different methods to calculate the failure probability of a structure using the load effect and structural information provided in Steps 2 and 3. This paper focuses on the exhaustive approach proposed by Lee et al., 2013a and Lee et al., 2013b, which consists of calculation of partial and total failure probabilities and normalization (i.e., converting the load effect distribution into a normal distribution).Lee et al., 2013a and Lee et al., 2013b proposed that, for two different time-variant loads, the failure probability of a structure is the sum of three partial failure probabilities from three mutually exclusive events. In this case, the total failure probability is

$$p_f = p_f^{EQ} + p_f^{EL} + p_f^{LL} \tag{1}$$

where p_f^{EQ}, p_f^{EL} and p_f^{LL} are the partial failure probabilities due to dead load and earthquake load effect, the combined effect of dead load, earthquake and live load, and dead load and live load effect only, respectively.

FORCE-BASED APPROACH AND DISPLACEMENT-BASED APPROACH TO EVALUATE BRIDGE RELIABILITY UNDER EARTHQUAKE

Force-based approach

Ghosn et al. (2003) proposed that the applied moment on the bridge column can be calculated using the following expression

$$M_{apl} = \lambda_{eq} C' S_a(t'T) \frac{AWH}{R} \tag{2}$$

where M_{apl} is the applied moment at the base of the column, λ_{eq} is the modeling factor for the analysis of earthquakeloads on bridges, C' is the response spectrum modeling parameter, A is the maximum 75-year PGA at the site (a 75-year design life is used to be consistent with AASHTO LRFD Specifications), S_a is the calculated spectral acceleration as a function of the actual period, T is the bridge column period, t' is the period modeling factor, W is the weight of the system, R is the response modification factor, H is the column height.

Eq. (2) takes many parameters and their uncertainties into account by calculating the distribution of earthquake load effect. Ghosn et al. (2003) assumed a significant parameter—response modification factor R follows a normal distribution with a mean value of 7.5 and a coefficient of variance (COV) of 34%. However, this assumption for Rmight not be reasonable for many cases. For an actual bridge column, the bending moment under lateral load can be simply depicted by idealized bi-linear model with two stages: elastic stage and inelastic stage. The load effects can not be modified in elastic stage while they have to be modified appropriately in inelastic stage. In other words, Eq.(2) may underestimate the bending moment in elastic stage and overestimate that in inelastic stage by using the same random response modification factor, as depicted in Fig. 2.

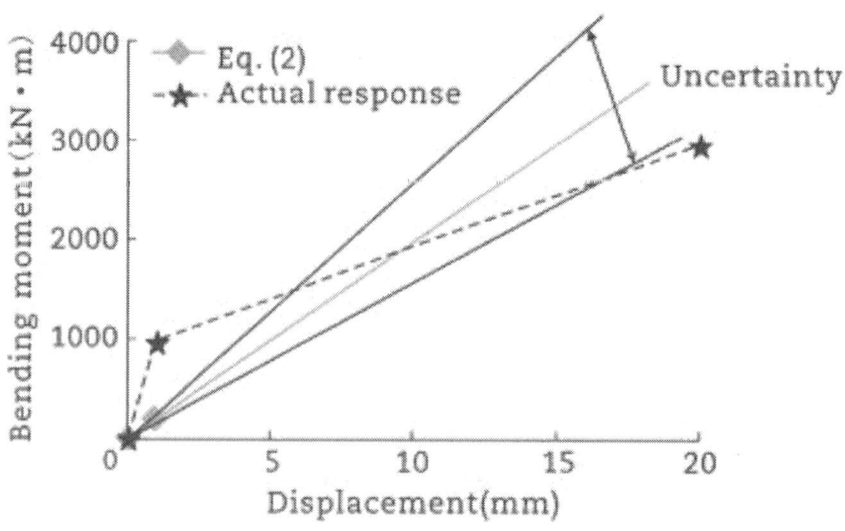

Figure 2.Comparison of actual column response and Ghosn's approach.

To address this issue, this paper proposes a different approach developed by Lee et al., 2013a and Lee et al., 2013b. Detailed formulations and calculations can be referred toLee et al., 2013a and Lee et al., 2013b. Since earthquake is the major cause for the failure of bridge substructure, this section focuses on introducing a force-based approach and a displacement-based approach to evaluate the bridge reliability under earthquake. To obtain an accurate distribution of earthquake load effects, a time-history analysis is used in this study. Several steps are as following:

1) Obtain the peak ground motion distribution for one event. According to the definition of Poisson process assumption and introduction in the later section, the distribution of earthquake PGA for one event can be obtained by

$$_{\mathrm{EV}}F_{\mathrm{PGA}}(x) = {}_{\mathrm{an}}F_{\mathrm{PGA}}(x)^{\frac{1}{n_{\mathrm{EQ}}}} \tag{3}$$

where $_{\mathrm{EV}}F_{\mathrm{PGA}}(x)$ and $_{\mathrm{an}}F_{\mathrm{PGA}}(x)$ are the cumulative distribution function (CDF) of PGA for earthquakes for one event and the annual CDF, respectively, n_{EQ} is the average number of earthquake occurring in one year.

The threshold PGA values for defining an earthquake is an important factor to determinen_{EQ}. Apparently this number of earthquakes will increase with the decrease of the threshold value.

Based on the database by United States Geology Survey (USGS), Fig. 3 shows the annual probability of exceedance for four different cities in U.S., including New York City, San Francisco, Seattle and St. Paul. The annual CDF of PGA for earthquake is $1-{}_{\mathrm{an}}F_{\mathrm{PGA}}(x)$.

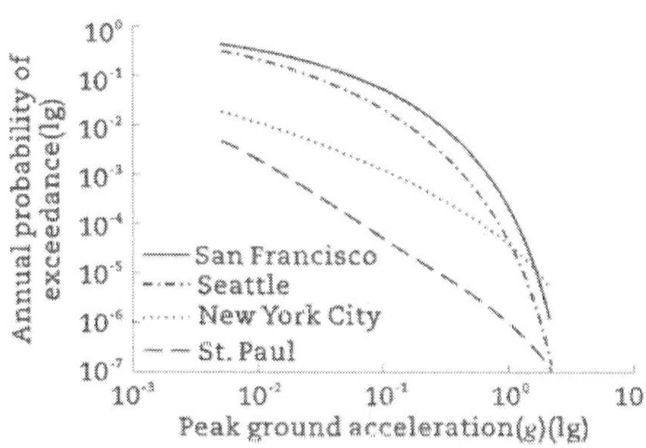

Fig. 3. Peak ground acceleration and corresponding annual probability of exceedance.

According to the statistics by USGS, the numbers of earthquakes in 75 years for the four cities above are 30, 600, 150 and 1, respectively (Ghosn et al., 2003).

2) Calculate the probability density function (PDF) for one earthquake based on their definitions by

$$ {}_{EV}f_{PGA}(x) = \frac{d}{dx}\,{}_{EV}F_{PGA}(x) \tag{4} $$

where ${}_{EV}f_{PGA}(x)$ is the PDF of PGA for earthquakes for one event. Numerically, the PDF of PGA is equivalent to the probability mass function (PMF).

3) Discretize the PDF for one event with a specified interval. In this study, the maximum considered earthquake level is 2.50g, and the PGA is divided into 251 levels with an interval of 0.01g (i.e., 0, 0.01g, 0.02g, 0.03g,⋯,2.50g).
4) Select the earthquake ground motions and match them with an objective response spectrum. A total of 99 earthquake ground motions are selected from Pacific Earthquake Engineering Research Center (PEER) next generation attenuation (NGA) database. With the objective to develop new ground-motion prediction relations through a comprehensive and highly interactive research program, PEER NGA provides robust and reliable estimates of ground motion than the previous generation of attenuation models developed in 1990s and early 2000s. To obtain a more accurate result, the selected ground motions are matched to an objective response spectrum provided by USGS for the selected location. Fig. 4 shows the response spectrum for the selected earthquake ground motions before and after matching process. The selected set of earthquake ground motion records is scaled to each level of PGA (i.e., 0.01g, 0.02g,0.03g,⋯,2.50g). The ground motions are linearly scaled by the PGA amplitude in this study.

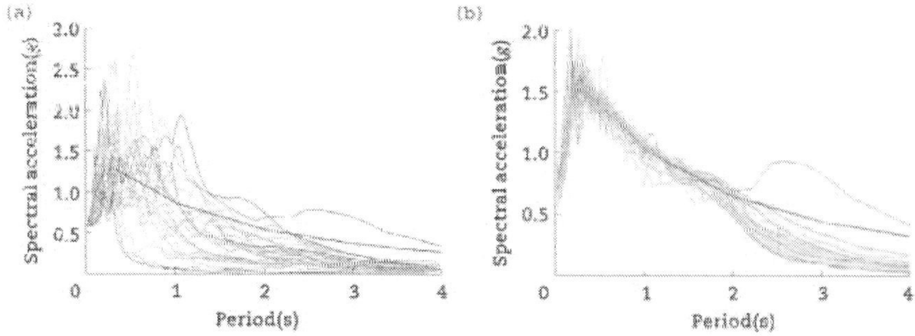

Fig. 4. Response spectrums of selected ground motions. (a) Response spectrums of ground motions before matching process. (b) Response spectrums of ground motions after matching process.

5) Convert the entire bridge models to a simplified model consisting of single or multiple columns. Priestley et al. (1996) proposed a model on simplifying the bridge bent to a simple bridge system, as shown in Fig. 5. In general, bridge substructures are more vulnerable to earthquake ground motions than superstructures. As mentioned earlier, this research is focused on the bending moment at the column base. Conducting the time-history analysis on a full bridge model is neither efficient nor necessary. Therefore, an equivalent spring to substitute the superstructure is developed to consider the inertial force resisted by the superstructure (Fig. 6).

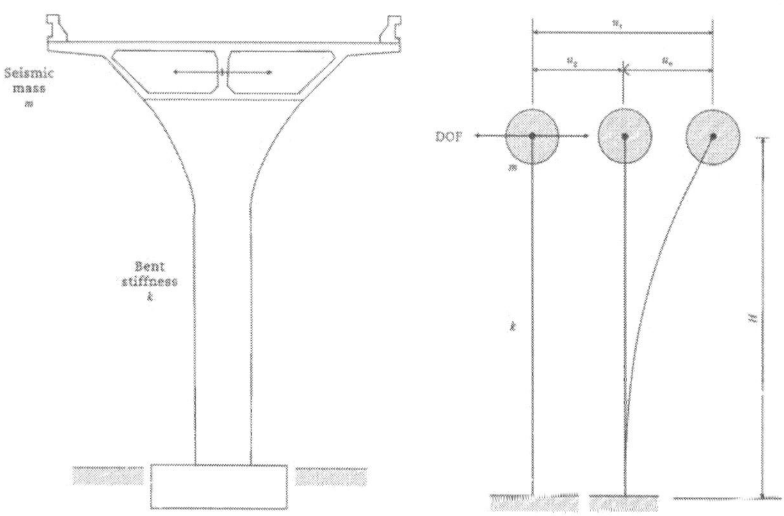

Fig. 5. Dynamic bridge response model.

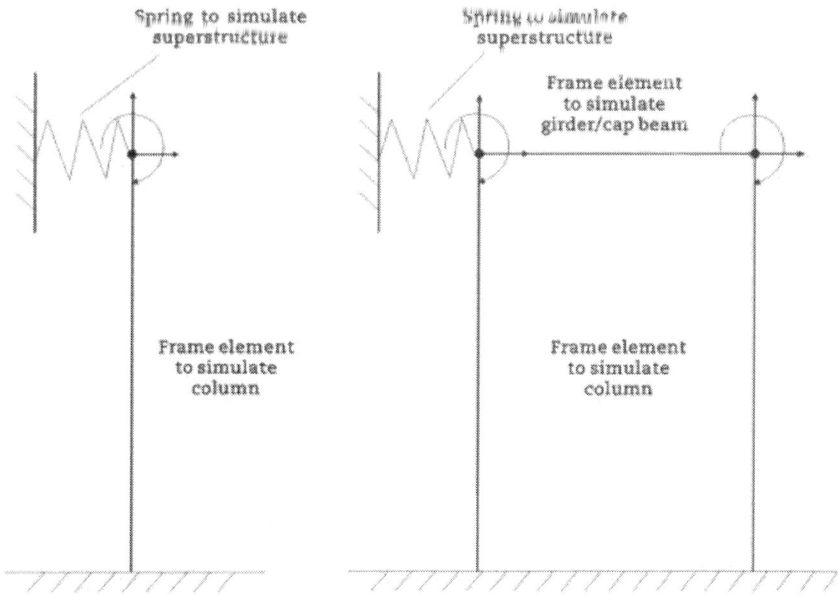

Fig. 6. Simplified model for bridge in longitudinal and lateral directions.

6) Determine the column property. In this study, the column section is 1.22 m in diameter and has 38 #11 bars for longitudinal steel, which amounts to a steel-concrete area ratio of about 1% (Fig. 7). In the plastic hinge zones, the columns have confinement steel consisting of #6 spiral bars with a 89 mm spacing. The concrete and steel used in the columns are 28 MPa and A706 of 420 MPa steel, respectively. The column is idealized by a bi-linear behavior, as shown in Fig. 8. d_m^+, d_y and d_m^- are maximum, yielding and minimum displacements, respectively, F_m^+, F_y and F_m^- are maximum, yielding and minimum forces, respectively, K_d, K_u and K_{eff} are loading, unloading and effective stiffnesses, respectively, Q_d is the characteristic strength (Hwang et al., 1996). The moment-curvature curve of the column in this study is depicted in Fig. 9. The axial load of the column has been considered in the moment-curvature curve.

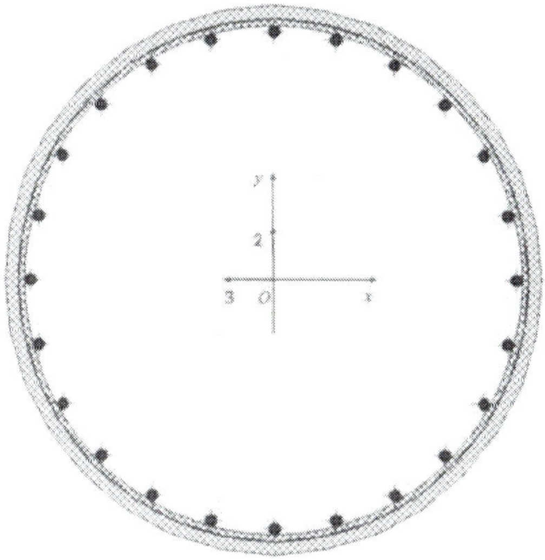

Fig. 7. Cross section of column.

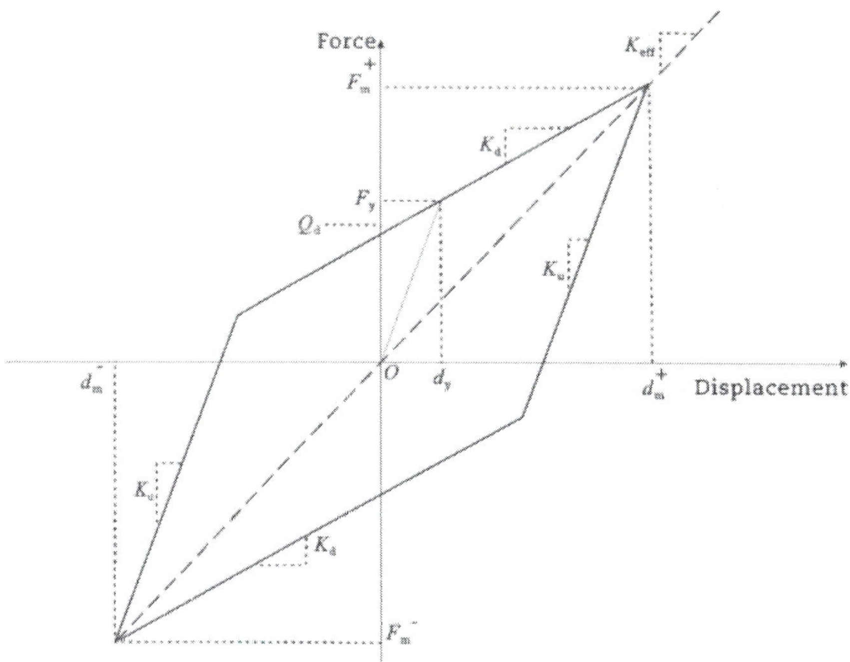

Fig. 8. Column property: idealized bi-linear hysteretic model.

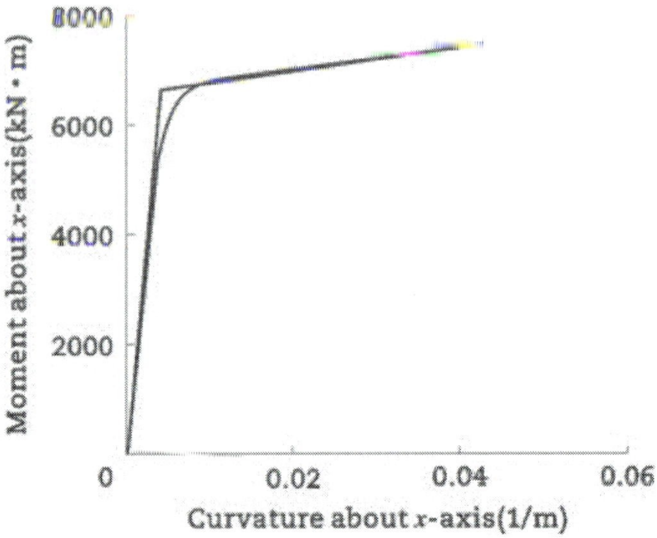

Fig. 9. Moment-curvature curve of column.

7) Analyze the time-history response of the column by each scaled ground motion, and then record the maximum load responses in the abscissa of the PMF. The corresponding probability of PGA is recorded for ordinate value as depicted in Fig. 10.

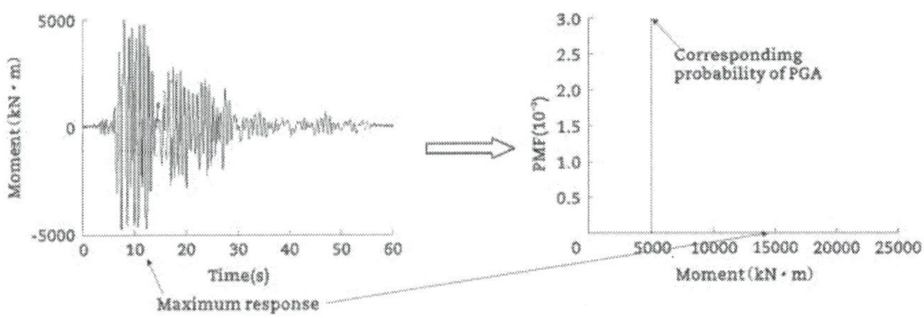

Fig. 10. Force-based time-history response of column.

8) Normalize the PMF to ensure the sum of probability equals 1. As a result, the PMF of earthquake load effect for one event can be obtained as depicted in Fig. 11(a).

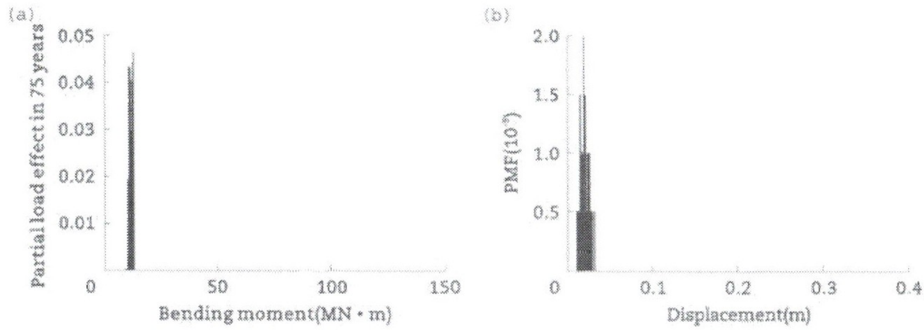

Fig. 11. Results of distribution of partial load effect. (a) Force-based approach. (b) Displacement-based approach.

Displacement-based approach

In general, the displacement-based approach follows the same steps as the force-based approach. They are:

1) Calculate the peak ground motion distribution for one event.
2) Compute the probability density function (PDF) for one earthquake.
3) Discretize the PDF for one event with a pre-set interval.
4) Select and match the earthquake ground motions using a targeted response spectrum.
5) Convert the bridge models to a simplified model that consists of single or multiple columns, depending on the column number in the original bent.
6) Compute the column property.
7) Analyze the time-history response of the column by each scaled ground motion, and then record the maximum load responses in the abscissa of PMF, the corresponding probability of PGA is recorded for ordinate value as depicted in Fig. 10.
8) Normalize the PMF to obtain the final distribution of earthquake hazard.

It is noted that in Step 7, the time-history responses of the column are selected as the displacement instead of bending moment. This difference may have a significant impact on the final results if the hardening stiffness of the column is relatively small. The change of loading force stays within a very limited range in this case. However, the displacement can vary in a larger scale because the hardening stiffness is always less than 1. Fig. 12shows the procedure in Step 7 and Fig. 11(b) shows the result. It can be seen that the variation of PDF based on displacement approach is larger than the force-based PDF (Fig. 11(a)).

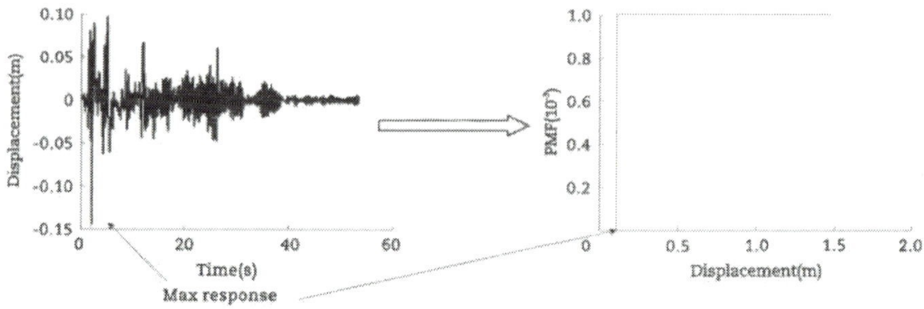

Figure 12. Displacement-based time-history response of column.

The other significant differences between force-based approach and displacement-based approach are the identification of resistance. For force-based approach, the mean value of resistance is set as the nominal yielding strength of column. For the displacement approach, the mean value of resistance is assumed to be the response reduction *R* factor regulated in AASHTO Specifications times the yielding displacement. In this case, the bridge *R* factor is selected as 3 for single-column bent and 5 for multiple-column bent.

Case study and comparison of two approaches

A set of nine simplified bridge models presented in a previous study by Lee et al., 2013a and Lee et al., 2013b has been used for the comparison of two approaches.

The main span length of the bridges varies from 21 m to 61 m so they are in the category of short-to-mid span typical bridges. Fig. 13 shows the 3D scheme of a typical bridge model, i.e., a three-span single-column pier bridge. The bridge is assumed to locate in San Francisco, CA with its column designed based on current AASHTO LRFD Specifications. Table 1 shows the detailed information about nine different bridge models, including the side and central span lengths, girder depth, column height, column diameter, longitudinal and transverse rebar details, etc.

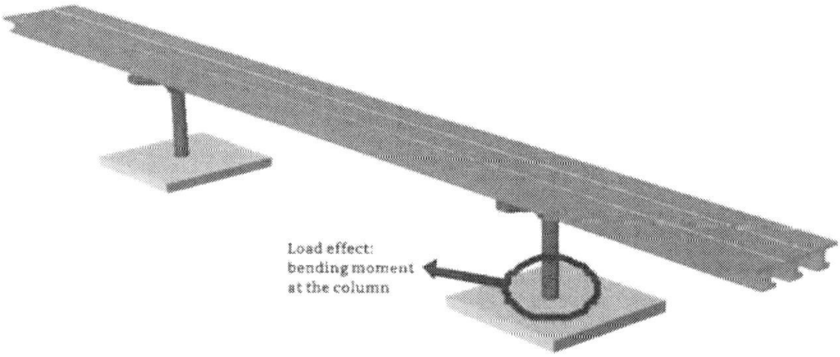

Figure 13. 3D scheme of a typical bridge model.

Table 1. Nine different bridge models presented by Lee et al., 2013a and Lee et al., 2013b.

Table 1 – Nine different bridge models presented by Lee et al. (2013).

Bridge no.	Span length (m)		Box girder height (m)	Column height (m)	Column seismic design result				
	Side span	Central span			Diameter (m)	Longitudinal rebar	Transverse rebar (spiral)		
							General (mm)	End (mm)	P. Hinge (mm)
1	30	36	1.8	6.0	1.2	(2)19-#11	#5 @55	#5 @44	#5 @85
2	18	21	1.8	6.0	1.2	22-#11	#5 @110	#5 @70	#5 @85
3	51	61	2.4	6.0	1.7	(2)26-#14	3#6 @110	3#6 @100	3#6 @225
4	30	36	1.8	15.0	1.7	(2)19-#11	#5 @200	#5 @100	#5 @ 60
5	18	21	1.8	15.0	1.7	33-#11	#5 @200	#5 @100	#5 @60
6	51	61	2.4	15.0	2.4	52-#14	#6 @110	#6 @50	#6 @50
7	30	36	1.8	10.5	1.7	30-#14	#5 @100	#5 @50	#5 @60
8	18	21	1.8	10.5	1.2	(2)21-#11	#5 @100	#5 @75	#5 @85
9	51	61	2.4	10.5	2.1	56-#14	2#6 @110	2#6 @85	2#6 @100

According to the procedure mentioned in the earlier sections, the reliabilities of these bridges under dead load, live load, and earthquake are calculated and the results are shown in Fig. 14. In this study, the response is analyzed by MATLAB Simulink to expedite the computation procedure. Note that the earthquake ground motions were all matched to design response spectrum regulated by AASHTO LRFD Specifications for San Francisco. The reliability obtained by the force-based approach varies from 1.27 to 11.54 and that by the displacement-based approach varies from 1.74 to 4.51.

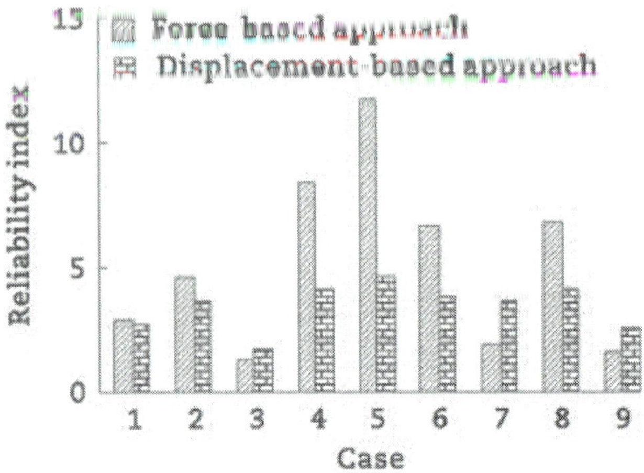

Figure 14. Comparison of two approaches.

In general, the results from two approaches show similar trends for the nine cases. The reliability of bridge increases with the increasing column height and the decreasing bridge span.

The displacement-based approach gives a lower reliability index. The reason can be attributed to two aspects: 1) the larger COV due to the more wide-spread distributed results from the nature of displacement-based approach; 2) the assumption of displacement resistance may be conservative since the response modification factor is selected by current specification. For cases 3 and 9, the large span lengths of the bridge lead to enormous inertia forces, which can be attributed to the large mass from bridge superstructure. The resistance of the force-based approach is relatively small thus it gives a lower reliability index in these two cases.

Generally, the result from the displacement-based approach is more consistent and stable for the nine cases selected. Considering the large uncertainty of natural earthquake, it can be concluded that the displacement-based approach provides less variations and thus comes up with more convergent and accurate reliabilities for these selected models.

CONCLUSIONS

This paper presents two approaches for the evaluation of the bridge reliabilities under the limit state of extreme event I combination of dead load, live load and earthquake load. The difference is on the step of converting load effect to load effect distribution. A set of nine simplified bridge models presented in previous study has been used for the comparison of two approaches. Results show that the displacement-based approach comes up with more convergent and accurate reliabilities for selected models. Note that the data and samples of bridge models are limited and further calibration are necessary to validate the accuracy of both approaches.

ACKNOWLEDGMENTS

The authors express their sincere appreciation to the financial support of FHWA (No.DTFH61-08-C-00012) with Dr. Phil W. Yen as the COTR. The paper is also funded by China Postdoctoral Science Foundation (No. 2012M511959), National Engineering and Research Center for Highways in Mountain Area (No. gsgzj-2012-07) and Applied Basic Research Project of MOT (No. 2015319812010). The writers gratefully acknowledge the provided support. The viewpoint of this paper does not reflect the sponsor's opinion.

REFERENCES

1. AASHTO, 2012. AASHTO LRFD Bridge Design Specifications, sixth ed. AASHTO, Washington DC.
2. Akiyama, M., Frangopol, D.M., Arai, M., et al., 2013. Reliability of bridges under tsunami hazards: emphasis on the 2011 Tohoku-Oki Earthquake. Earthquake Spectra 29 (S1), 295e314.
3. Borges, J.F., Castanheta, M., 1972. Statistical Definition and Combination of Loads, in Probabilistic Design of Reinforced Concrete Buildings. American Concrete Institute, Detroit.
4. Cui, P., Wei, F., He, S., et al., 2008. Mountain disasters induced by the earthquake of may 12 in Wenchuan and the disasters mitigation. Journal of Mountain Science 26 (3), 280e282.
5. Fujikura, S., Bruneau, M., 2012. Dynamic analysis of multihazardresistant bridge piers having concrete-filled steel tube under blast loading. Journal of Bridge Engineering 17 (2), 249e258.
6. Ghosn, M., Moses, F., Wang, J., 2003. Design of Highway Bridges for Extreme Events. NCHRP Report 489. Transportation Research Board, Washington DC.

7. Huang, C., Shen, J.J., Liang, Z., 2014. Comparison on force-based and displacement-based approaches for evaluation of bridge reliability under multiple hazards. In: 10th Asia Pacific Transportation Development Conference, Beijing, 2014.
8. Hwang, J.S., Lin, C.Y., Ku, S.W., et al., 1996. Analysis accuracy implied by equivalent linear models of bridge isolation. In: 11th World Conference on Earthquake Engineering, Oxford, 1996.
9. Kameshwar, S., Padgett, J.E., 2014a. Multi-hazard risk assessment of highway bridges subjected to earthquake and hurricane hazards. Engineering Structures 78 (1), 154e166.
10. Kameshwar, S., Padgett, J.E., 2014b. Towards risk-based, multihazard-resistant design of bridges. In: Structures Congress 2014, Boston, 2014.
11. Kulicki, J.M., 1998. Development of Comprehensive Bridge Specifications and Commentary. National Cooperative Highway Research Program Research Results Digest No. 198. Transportation Research Board, Washington DC.
12. Kulicki, J.M., Prucz, Z., Clancy, C.M., 2007. Updating the Calibration Report for AASHTO LRFD Code. NCHRP Project 20-07/186. Transportation Research Board, Washington DC.
13. Lee, G.C., Huang, C., Liang, Z., et al., 2014. Towards next generation LRFD extreme event design limit states. In: Structures Congress 2014, Boston, 2014.
14. Lee, G.C., Liang, Z., Shen, J.J., et al., 2011. Extreme Load Combinations: a Survey of State Bridge Engineers. MCEER- 11-0007. Multidisciplinary Center for Earthquake Engineering Research, Buffalo.
15. Lee, G.C., Mohan, S.B., Huang, C., et al., 2013a. A Study of U.S. Bridge Failures (1980e2012). MCEER-13-0008. Multidisciplinary Center for Earthquake Engineering Research, Buffalo.
16. Lee, G.C., Tong, M., Yen, W.P., 2008. Design of Highway Bridges against Extreme Hazard Events. MCEER-08-SP06. Multidisciplinary Center for Earthquake Engineering
17. Research, Buffalo.
18. Lee, G.C., Tong, M., Yen, W.P., et al., 2013b. Application of the Principles and Approaches of Multiple-hazard Design for Highway Bridges: a Case Study on Dead Load, Live Load and Earthquake. Technical Report MCEER 13-00012. Multidisciplinary Center for Earthquake Engineering Research, Buffalo.
19. Li, Y., Ahuja, A., Padgett, J., 2012. Review of methods to assess, design for, and mitigate multiple hazards. Journal of Perform Constructed Facilities 26 (1), 104e117.
20. Liang, Z., Lee, G.C., 2012. Towards multiple hazard resilient bridges: a methodology for modeling frequent and infrequent time-varying loads. Journal of Earthquake Engineering and Engineering Vibration 11 (3), 303e311.
21. Liang, Z., Lee, G.C., 2013. Towards establishing practical multihazard bridge design limit states. Journal of Earthquake Engineering and Engineering Vibration 12 (3), 333e340.

22. Nowak, A.S., 1999. Calibration of LRFD Bridge Design Code. NCHRP Report 368. Transportation Research Board, Washington DC. Petrini, F., Palmeri, A., 2012. Performance-based design of bridge structures subjected to multiple hazards: a review. In: Proceedings of the 6th Internation IABMAS Conference, Stresa, 2012.
23. Potra, F., Simiu, E., 2009. Multihazard design: structural optimization approach. Journal of Optimization Theory and Applications 144 (1), 120e136.
24. Priestley, M.J.N., Seible, F., Calvi, G.M., 1996. Seismic Design and Retrofit of Bridges. John Wiley &Sons, Inc., New York.
25. Rini, D., Lamont, S., 2008. Performance based structural fire engineering for modern building design. In: Structures Congress 2008, Vancouver, 2008.
26. Sun, D.Z., Chen, B., Sun, B.T., 2015. Study based on bridge health monitoring system on multihazard load combinations of earthquake and truck loads for bridge design in the southeast coastal areas of China. Shock and Vibration 2015, 1e12.
27. Turkstra, C.J., Madsen, H., 1980. Load combinations in codified structural design. Journal of the Structural Division 106 (12), 2527e2543.

CITATION

Chao Huang, Jerry J. Shen, Mi Zhou, George C. Lee, Force-based and displacement-based reliability assessment approaches for highway bridges under multiple hazard actions, Journal of Traffic and Transportation Engineering (English Edition), Volume 2, Issue 4, August 2015, Pages 223-232, ISSN 2095-7564, http://dx.doi.org/10.1016/j.jtte.2015.05.002.

CHAPTER 7

Traffic Flow Consideration in Design of Freight Distribution System

Sutanto Soehodho, Nahry

Dept. of Civil Engineering, Faculty of Engineering, University of Indonesia, Kampus UI Depok 16424, Indonesia

ABSTRACT

This study is part of a series of research projects on a distribution system we developed to deal with cases in a state-owned company. It concerns the design of the Public Service Obligation State-owned Company (PSO-SOC) distribution system. The intrinsic features of PSO-SOC are distributing strategic commodities and having subsidies within the cost function. Hence their distribution flow has to be secured under consideration of moving the commodities within road networks that have traffic flow dependency. This paper focuses on the solution of the proposed model which represents traffic flow dependency within a freight distribution network.

The mathematical formulation takes the form of a Minimum Cost Multicommodity Flow (MCMF) problem. Traffic flow dependency is incorporated into the model by introducing a coefficient of speed, which is derived from the traffic assignment of ordinary traffic associated with the transportation of the type of freight under consideration The solution of the proposed model is formulated by Network Representation (NR), in which all of the components of the mathematical model are represented in the form of dummy links and nodes added to the original (physical) network. It is to be noted then, that the traffic flow on each road or link is represented by a link performance function (LPF), depicting traffic flow dependent travel time and consequent cost. The MCMF problem of NR is further solved by a Primal–Dual Algorithm.

Finally, an illustrative example is exercised to show how the proposed step-wise solution works.

INTRODUCTION

Research on designing freight distribution systems has been done for many years. Most deal with a private company whose primary concern is merely profit maximization. In Indonesia, there exists a subsidized service by State-Owned Company (SOC) called Public Service Obligation (PSO)-SOC. PSO-SOC has the obligation to serve the entire demand on public commodities or services. Its working orientation is not for profit, but for security of supply. The PSO-SOC is still permitted to conduct its own programs beyond its main task, but it is undertaken within government control and limitation. The PSO-SOC also bears strategic commodities in distribution, so availability of commodities at the right time and place is important. Most of those commodities are being transported through surface land transport with mixed-traffic, so travel time and cost are very much dependent on the flow of vehicles within the road network.

The identification of the distribution system of one of Indonesia's PSO-SOCs, which deals with the production and distribution of public commodities, provides insight to some important issues as follows [1]:

1. The company under consideration is a group of companies that consists of 1 (one) holding company and 5 (five) affiliated companies. Each of the companies (included the holding company itself) carries out the operation of its own plant and its distribution process independently. Those companies are managed separately and there is no regulation that integrates those companies in their logistical process.
2. Unit production costs are not uniform among the plants. This is due to the different prices of raw material and the variability in the operation performance of the production processes.
3. Product differentiation exists. This implies that products are not differentiated merely by type (material) of product, but also by the type of user. There are two types of user, public (subsidized) and commercial. Both of them are different in terms of selling price and demand satisfaction. Subsidized prices are determined by the government, while the commercial ones are set by the company. Naturally, the commercial prices are higher than the subsidized ones. Moreover, subsidized demands have the privilege of being fully satisfied regardless of the amount of profit that the company may receive from them.

Previous research on freight distribution systems was concerned mostly with private companies [2], [3] and [4]. Most of the research on the distribution of public needs is related to public services (such as schools, police stations, hospitals, etc.) rather than public goods. Savas [5] focused his research on the equity in providing public services, while Ross [6] proposed a model with multicriteria to select sites for public facilities. Regarding the variables included, most of the research focused on transportation costs, while some dealt with various other variables, such as production cost, fixed cost of facility, inventory holding cost, and anything relevant to the special problem they faced[2], [7], [8] and [9].

Moreover, most of the research on freight distribution systems hardly consider the effect of ordinary traffic movement on the system. The FHWA reports [10] that the private and public sectors of the freight industry bear an operational cost of congestion which is as much as $25–200 an hour, depending on product type and other factors. The report also estimates that unexpected delays can increase the cost of transporting goods by 50–250%. Jones [11] explains that for the freight industry and trucking companies, congestion on the transportation network diminishes productivity and increases the overall cost of transportation services significantly. Other effects of traffic delays are a higher cost for fleet operations, decreased fleet and vehicle utilization, decreased fuel efficiency, increased emissions due to idling, and decreased hours of "productive" service for drivers.

These intrinsic characteristics clarify the need to enhance the earlier distribution models for the purpose of taking into account the special role of a state-owned company and considering the movement of the commodities within road networks that have traffic flow dependency.

MODEL FORMULATION

In order to take into account the characteristics of the PSO-SOC as stated in Section 1, we propose that the distribution system of the affiliated companies should be integrated into one system which is coordinated by the holding company. Furthermore, we propose a distribution model which deals with production cost, transportation cost, warehouse cost, as well as revenue as its variables [1]. These variables are considered for the following reasons. When the total plant capacity is more than the total demand, the holding company has to designate production allocation to each plant, and in such a situation, one or more plants must be under capacity (not operated in full capacity). When the holding company only considers the distribution cost, it may lose efficiency on logistics as a whole, particularly if the cost of production is not in accordance with the efficiency at the distribution cost level.

In the case of over demand, all of the plants must be fully operated. In such case, the holding company must decide which commercial demands should be satisfied in order to attain the maximum profit of the company. Since the selling price of commercial products varies significantly, the selling price becomes a very important variable to be included in this optimization.

Regarding ordinary traffic considerations, we propose that a coefficient of speed be included in the model. The coefficients of speed are derived from ordinary traffic assignments which produce user-equilibrium link travel time and the associated speeds. It is actually quite valuable to consider the effect of ordinary traffic on the movement of freight vehicles, since it will lead to more informed and efficient decisions for freight allocation. Indeed, the decision to assign some amount of products to a set of plants and distribute them to a set of end consumers through certain paths is crucial to maintaining low transportation costs since these costs are highly affected by traffic performance on any given path.

The PSO-SOC under consideration takes into account only transportation costs and it does not consider the ordinary traffic flow effect in its product allocation process. Besides, the process is carried out by each plant independently (each plant is operated exclusively by one affiliated company) and there is no integration with the holding company.

In order to be more precise in making decisions on product allocation, we propose a model which deals with the cost of producing, transporting, and handling the commodities [1], as well as revenue, and includes the effect of ordinary traffic on the product allocation decision. Product allocation is also proposed to be optimized by integrating logistic subsystems of all plants.

In order to cope with the problem of a distribution system characterized mainly by product/demand differentiation, integrated systems, as well as traffic flow dependency, we propose a mathematical model which is applicable to the following distribution network, as depicted in Fig. 1. It consists of a set of plants, consolidation centers and retailers.

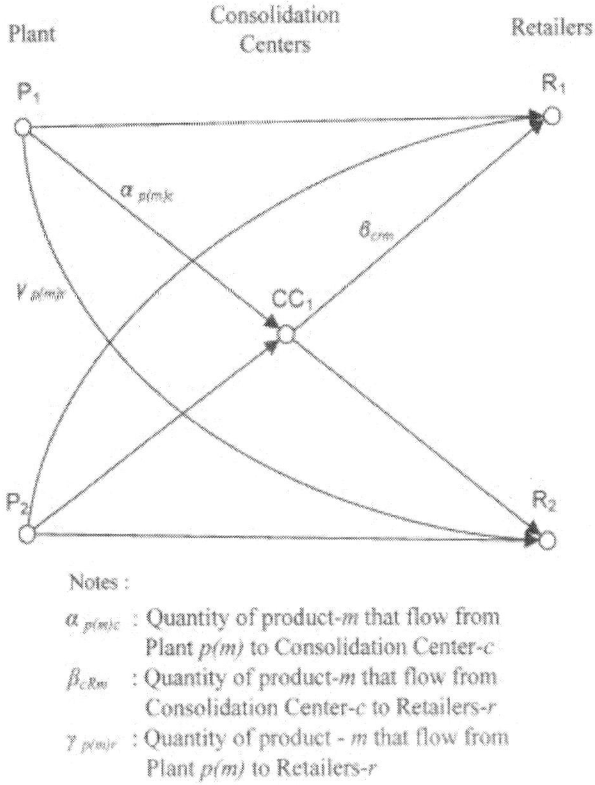

Figure 1. An example of a physical distribution network.

The proposed mathematical model is as follows:

Equation (1)

$$
\begin{aligned}
\min Z(\alpha_{p(m)c}, \beta_{crm}, \gamma_{p(m)r}) = & \sum_{p\in P}\sum_{c\in C}\sum_{m\in M} \mu_{pc} \cdot d_{pc} \cdot u_{pcm} \cdot \alpha_{p(m)c} \\
& + \sum_{c\in C}\sum_{r\in R}\sum_{m\in M} \mu_{cr} \cdot d_{cr} \cdot v_{crm} \cdot \beta_{crm} \\
& + \sum_{p\in P}\sum_{r\in R}\sum_{m\in M} \mu_{pr} \cdot d_{pr} \cdot z_{prm} \cdot \gamma_{p(m)r} \\
& + \sum_{p\in P}\sum_{c\in C}\sum_{m\in M} w_{cm} \cdot \alpha_{p(m)c} \sum_{p\in P}\sum_{m\in M} \\
& \times \left(\sum_{c\in C} \alpha_{p(m)c} + \sum_{r\in R} \gamma_{p(m)r} \right) \cdot \eta_{p(m)} \\
& - \sum_{r\in R}\sum_{m\in M} \left(\sum_{c\in C} \beta_{crm} + \sum_{p\in P} \gamma_{p(m)r} \right) \cdot \rho_{rm}
\end{aligned}
$$

Turn MathJax on

Equation (2)

$$\mu_{ij} = 1 - \frac{v_{UEij} - v_D}{v_D}$$

Turn MathJax on

subject to

Equation (3)

$$\sum_{p \in P} \alpha_{p(m)c} = \sum_{r \in R} \beta_{crm}, \qquad \forall c \in C, \forall m \in M$$

Turn MathJax on

Equation (4)

$$\sum_{c \in C} \beta_{crm} + \sum_{p \in P} \gamma_{p(m)r} = \lambda_{rm}, \qquad \forall r \in R, \forall m \in M_s$$

Turn MathJax on

Equation (5)

$$\sum_{c \in C} \beta_{crm} + \sum_{p \in P} \gamma_{p(m)r} \leq \lambda_{rm}, \qquad \forall r \in R, \forall m \in M_c$$

Turn MathJax on

Equation (6)

$$\sum_{c \in C} \sum_{m \in M} \alpha_{p(m)c} + \sum_{r \in R} \sum_{m \in M} \gamma_{p(m)r} \leq Cp_{p(m)} \qquad \forall p \in P, \forall m \in M$$

Turn MathJax on

Equation (7)

$$\alpha_{p(m)c} \geq 0, \qquad \forall p \in P, \forall c \in C, \forall m \in M$$

Turn MathJax on

Equation (8)

$$\beta_{crm} \geq 0, \qquad \forall c \in C, \forall r \in R, \forall m \in M$$

Turn MathJax on

Equation (9)

$$\gamma_{pcmr} \geq 0, \qquad \forall p \in P, \forall r \in R, \forall m \in M$$

Turn MathJax on

Subscripts

p:

indicate the plants

c:

indicate the consolidation centers

r:

indicate the retailers

m:

indicate the products

$p(m)$:

indicate the plant $p \in P$ that produces product-m

Sets

P:

set of plants

C:

set of consolidation centers

R:

set of retailers

M:

set of products

$Ms \in M$:

set of subsidy (public) products

$Mc \in M$:

set of commercial products

Decision variables:

$\alpha p_{(m)}c$

is quantity of product-*m* that flow from plant *p*(*m*) to consolidation center-*c*

βcrm

is quantity of product-*m* that flow from consolidation center-*c* to retailer-*r*

$\gamma p_{(m)}r$

is quantity of product-*m* that flow from plant *p*(*m*) to retailer-*r*

Input parameters

$upcm$:

per-mile cost to ship a unit of product-*m* from plant-*p* to consolidation center-*c*

$vcrm$:

per-mile cost to ship a unit of product-*m* from consolidation center-*c* to retailer-*r*

$zprm$:

per-mile cost to ship a unit of product-*m* from plant *p*(*m*) to retailer-*r*

d_{ij}:

length of distance of link *i*–*j*

μ_{ij}:

coefficient of speed of link *i*–*j*

$vUEij$:

user-equilibrium speed of link *i*–*j*

wcm:

unit warehouse cost to handle product-*m* in consolidation center-*c*

$\eta p(m)$:

unit cost for producing product-*m* in plant-*p*

prm:

selling price of product-*m* at retailer-*r*

$Cpp(m)$:

capacity of plant-*p* to produce product-*m*

demand of product-*m* in retailer-*r*

design speed of link *i*–*j*

Eq. (1) denotes the objective function of the proposed model. It actually maximizes the profit, in which profit is represented by revenue minus cost. Similarly, this objective function can be replaced by a minimization of cost function, which is represented by cost minus revenue.

The first three terms of Eq. (1) represent transportation cost, in which each term includes a coefficient of speed. Each coefficient of speed is exclusive for a certain link. It is derived from the user-equilibrium speed of an ordinary traffic assignment and design speed, as formulated in Eq. (2). The use of coefficients in Eq. (1) is meant to indirectly make corrections to the design unit cost through a correction of the design speed due to the dynamics of traffic conditions in real life situations. Since traffic conditions are very dynamic in nature, the time windows considered should be carefully selected. Moreover, this coefficient indicates that the smaller the speed (the more congested the road) the longer the "distance" that the freight vehicle should travel. The fourth term of Eq. (1) is related to warehouse cost. The fifth term represents production cost and the last term concerns revenue. Obviously, due to the opposite characteristics of cost and revenue, we must put a minus sign before revenue. Eq. (3) denotes that the total inflow minus the total outflow in consolidation centers is set at zero since those nodes are intermediate nodes.

Eqs. (4) and (5) are related to demand satisfaction for subsidized products and commercial ones, respectively. Subsidized products must be entirely fulfilled, while the commercial ones could be satisfied later, in the case of excess plant capacity. Eq. (6)implies that the total amount of production of any product by each plant should not be more than its capacity. Eqs. (7), (8) and (9) ensure non negativity of flow constraints.

Model solution

Since the MCMF problem is highly related to the network structure, we utilize Network Representation (NR) to represent and solve the proposed mathematical model. Network Representation (NR) is a technique used to solve a problem by representing a mathematical model as a network flow-based formulation [12]. It is characterized by the use of diagrams that have emerged, by progressive expansion, from those used traditionally in network flow and graph theory. Network Representation is developed by adding some dummy links and nodes into the original (physical) network, in which the function of those dummy links is designated to represent production cost, transportation cost, and warehouse cost, as well as revenue.

One particular issue regarding our MCMF problem is the involvement of multi-commodities in the distribution system. This issue is solved by introducing a Sub-Network Representation devoted to a certain product. We name such sub-NRs as Product Sub-Network Representation (P-SNR). Each P-SNR is exclusively devoted to a certain product, and consists of nodes and links used by that certain product, although it is possible that some of the links of a P-SNR are used for common products.

In order to cope with the situation of an imbalance between total supply and total demand, we developed an Excess Supply/Demand Sub-Network Representation. Its function is to balance the total supply and total demand endogenously in the optimization process, by decreasing the demand or supply so that both of them finally are in balance.

An example of a general NR is shown in Fig. 2. It consists of 2 plants, 1 consolidation center and 2 retailers, and it deals with 2 types of products and 2 types of demand (subsidized and commercial demand). The P-SNRs of such an NR are shown in Fig. 3a and b.

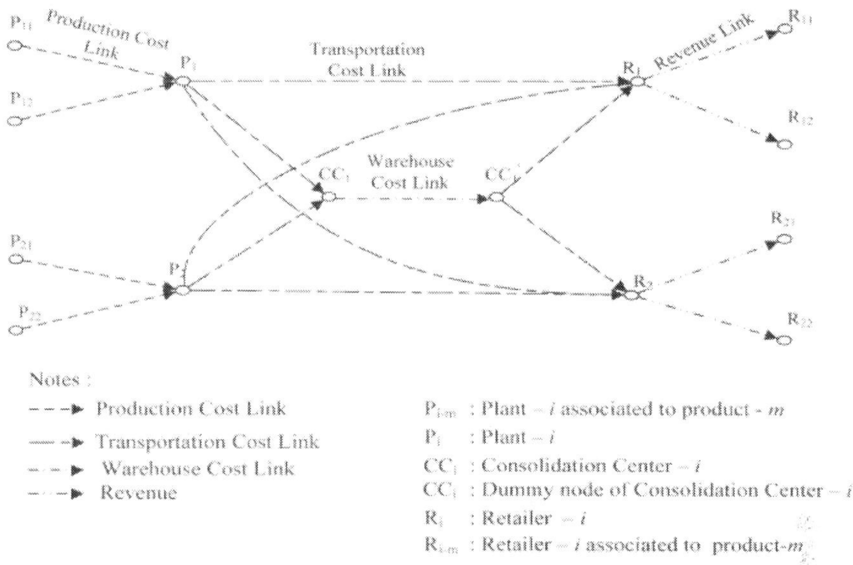

Figure 2. An example of a Network Representation.

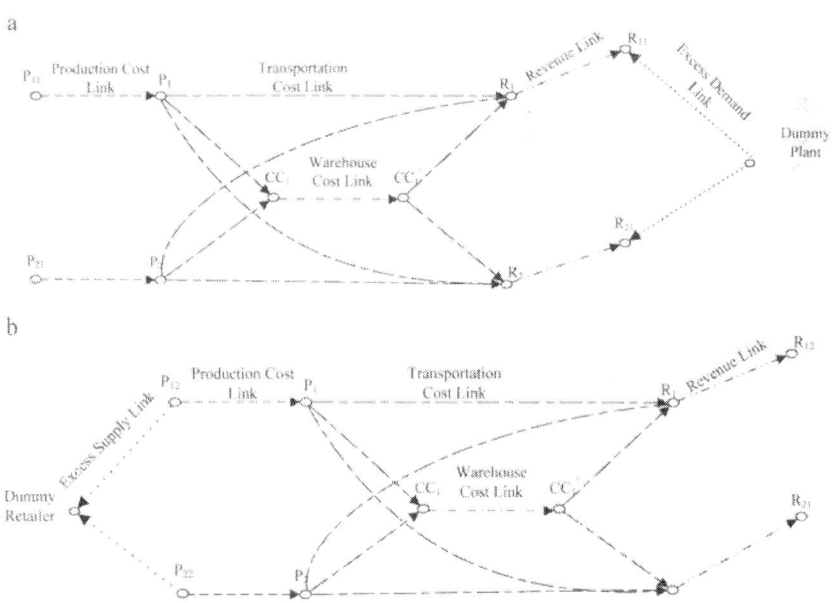

Figure 3. a. Product Sub-Network Representation of product-1 (in Excess Demand case). **b**. Product Sub-Network Representation of product-2 (in Excess Supply case).

Links between node *Pi–m* and *Pi* are designed as production cost links. Each link represents the cost to produce product-*m* in plant-*i*. Each of those links is also characterized as a product-exclusive link; that is each link is devoted to a certain product. Hence, in Fig. 3a, which shows the P-SNR of product-1, there are only nodes and links that relate to product-1 (link P_{11}–P_1, P_{21}–P_2).

Links between *Pi–CCi*, *Pi–Ri* and *CCi'–Ri* are designed as transportation cost links, and they represent the transportation cost between two distribution facilities. It is assumed that the unit cost to transport any type of product in a certain link is similar. Coefficients of speed are employed in each transportation cost link. The link between *CCi–CCi'*represents the cost of using warehouse-*i*. Every one unit of flow that comes to the consolidation center is charged by one unit of warehouse cost. Links between node *Ri* and *Ri–m* are designed as revenue links. Those links represent revenue from selling product-*m* to retailer-*r*. Each revenue link is also designed as a product-exclusive link. Unit cost associated to revenue link is denoted by selling price. All of the links of the NR are designed as un-capacitated links; hence there will be as many flows as possible passing through the links.

In a real situation, the total supply is not always in balance with the total demand. In some cases, the total supply is higher than the total demand. In such excess supply cases, the company should be selective in making product assignments to each plant. In excess demand cases, in which the total demand is higher than the total supply, the company should be selective in demand satisfaction. In the case of the PSO-SOC, subsidized demand should have priority to be fulfilled, no matter what the profit that the company could attain from it. Regarding both cases, we developed an Excess Supply/Demand Sub-Network Representation in order to accommodate both situations and to make the supply and demand be always in balance during the optimization process. Fig. 3a depicts the NR in the excess demand case, in which we added a Dummy Plant to act as a supply for the excess demand. Fig. 3b depicts the excess supply case, in which we added a Dummy Retailer to act as a receiver for some excess products from nodes of plants.

In order to give priority to the subsidized demand, we designate an extremely high unit cost to the links of the excess Demand Sub-Network that relates to nodes of subsidized (public) demand. This means that such a high "unsatisfied-demand cost" will cause the model to avoid unfulfillment of public demand

The nodes of the NR are valued as their flow requirement. Node *Pi_m* is valued by the capacity of plant-*i* on producing product-*m*, meanwhile node *Ri_m* is valued by the demand on product-*m* in retailer-*i*. The flow requirements of the intermediate nodes are set as zero.

Having the NR formulation, the problem now is how to determine the optimal assignment of the products to attain the "minimum cost". The minimum cost flow problem is solved by a Primal–Dual Algorithm [13]. We propose a step-wise solution of the MCMF problem as depicted in Fig. 4. It can be explained as follows:

Step 1: The process is initialized by doing traffic assignments to the ordinary traffic of the physical network of the freight distribution system. Provide the attributes of the ordinary traffic network, including link capacity, link performance function and O/D flow. Set the time window and do traffic assignments to find the user-equilibrium link travel time and the associated speed. Provide the Coefficient of Speed of each link using Eq. (2).

Step 2: Develop the General Network Representation (NR) of the freight allocation problem. Define the link capacity and the flow requirement as well as the unit cost of the NR by considering the Coefficient of Speed of step 1.

Step 3: Develop product sub-NR (P-SNR) of all types of products.

Step 4: Add an Excess Supply/Demand Sub-Network to the P-SNR of step 3; if it is needed, include setting of its link capacity and unit cost.

Step 5: Solve the MCF problem of the P-SNR by using a Primal–Dual Algorithm.

Step 6: Find the optimal flow and its associated paths through the P-SNR. Steps 3–6 are repeated until all products are optimized. Since one P-SNR is independent to another P-SNR, steps 3–6 are actually able to be carried out in a simultaneous way.

Step 7: When all the P-SNRs have been optimized, superimpose the optimal paths of all P-SNRs and find the total optimal flows on each link of the NR. Superimposing is allowable since it is assumed that unit cost to transport and handle any type of product in a certain link is similar, and each link of production cost, as well as revenue links, is characterized by a product-exclusive link.

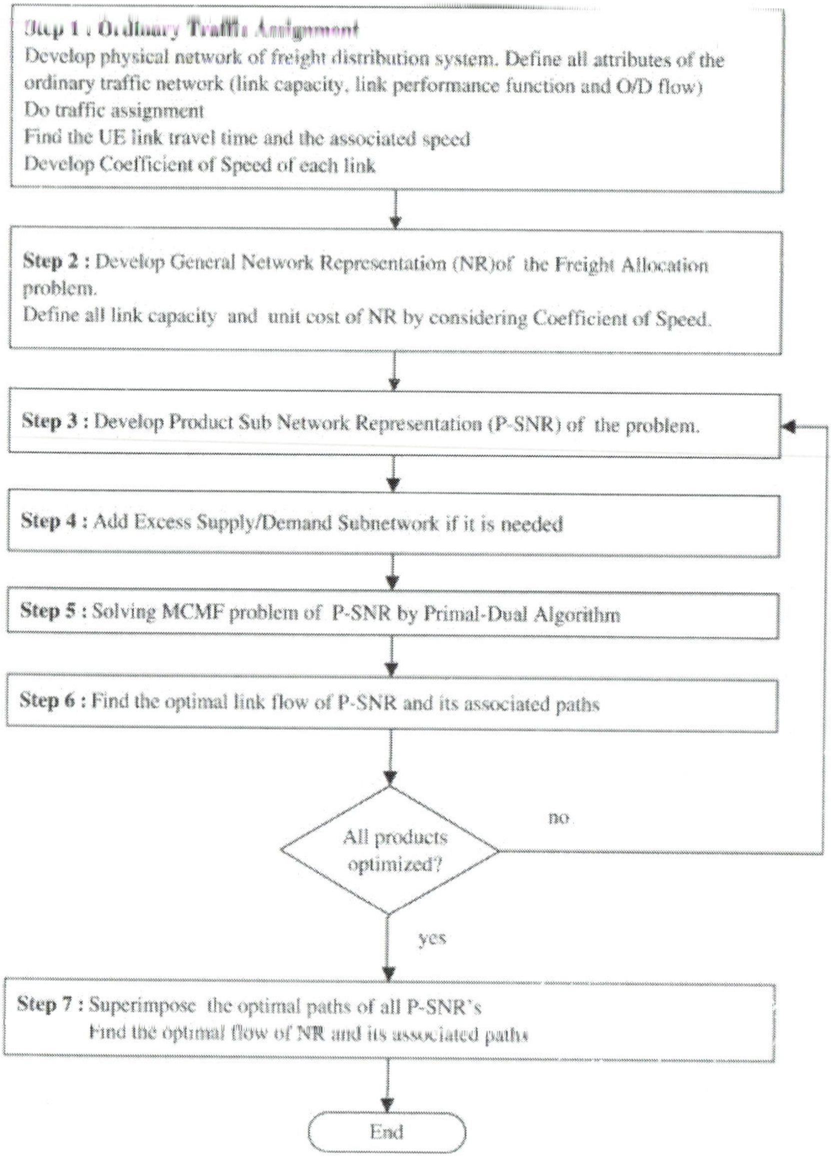

Figure 4. Step-wise of model solution.

Illustrative example

In an attempt to apply the step-wise proposed in Section 4, the ensuing contrived example is discussed. The distribution network of the example consists of 2 plants, 1 consolidation center and 2 retailers. It deals with 2 kinds of products and 2 kinds of demand (subsidy and commercial). Plant capacity and demand on each product are shown in Table 1 and Table 2 respectively.

Table 1. Plant capacity.

Plant	Plant capacity (units)	
	Product-1	Product-2
1	140	100
2	210	660
Total	350	760

Table 2. Demand on certain product.

Retailer	Demand on product (units)			
	1^S	1^C	2^S	2^C
1	50	20	10	30
2	30	10	20	40
Total	110		100	

Notes: s: subsidy; c: commercial.

We exercised this example with 3 cases. In the first case, it is assumed that coefficients of speed of all links are one. In the second and third cases, it is assumed that step 1 of the step-wise model is already done and 2 sets of link coefficients of speed were found. The first set is derived from a traffic assignment which is based on user-equilibrium with an average speed of 58.63 units (case 2) and the second one is based on user-equilibrium with an average speed of 61.2 units (case 3).

The optimal flow of the three traffic assignments is shown in Fig. 5 and the associated product assignment is displayed in Table 3. The values of the objective function, as well as the patterns of product assignment of all cases, are changed as the coefficient of speed changes. It can be concluded that the coefficients of speed are sensitive to the product assignment, as well as the value of the objective function. Moreover, it can be said that ordinary traffic flow is a necessary factor to be considered in designing a freight distribution system model in order to make it more realistic.

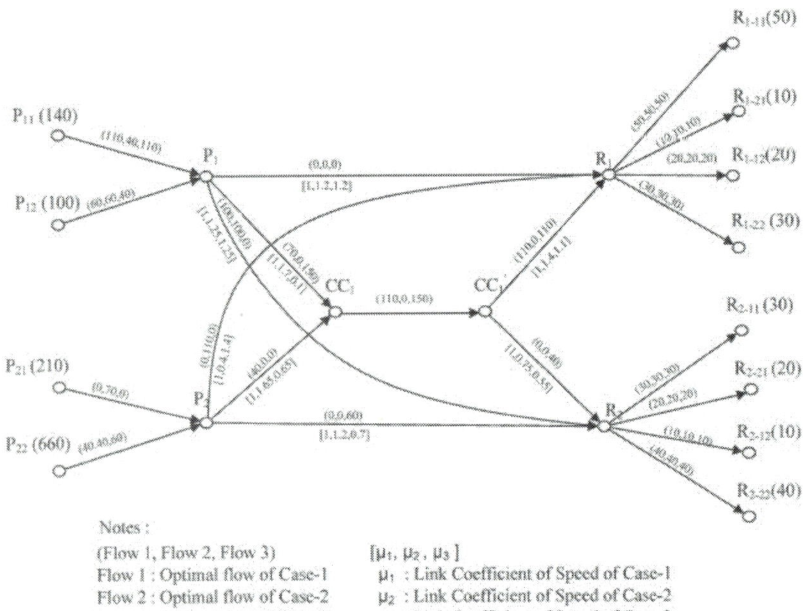

Figure 5. The optimal flows of 3 (three) distribution assignments.

Table 3. The assignment of the products.

Retailer	Product	Supplied by plant Case 1 Objective value = − 2670	Case 2 Objective value = − 2538	Case 3 Objective value = − 2896
1	1^S	1	2	1
2	1^S	1	1	1
2	1^C	1	1	1
1	1^C	1	2	1
1	2^S	2	2	1
2	2^S	1	1	2
2	2^C	1	1	2
1	2^C	2	2	1
Average speed		60 (Design speed)	58.63	61.2

Notes: s: subsidy; c: commercial.

CONCLUSION

We propose an allocation model which considers traffic flow dependence and the characteristics of the PSO-SOC. Traffic flow, that may increase travel cost through LPF and hence distribution cost, is accounted for within the traffic assignment. By doing so, the flow of commodities within mixed-traffic will implicitly represent cost performance due to the nature of traffic flow situations. The proposed model takes the form of a

Minimum Cost Multicommodity Flow problem and the solution is formulated by Network Representation.

Calibration of the model may be required when dealing with real cases in determining the coefficients of speed. The speed coefficient given in the example is a simple one to represent the relation between traffic flow patterns and the pattern of distribution. A more realistic one, which is close to a real condition representation, may be made with an empirical process through model calibration. The three cases given in exercising the model merely denote the different distribution patterns due to different traffic patterns. Those three cases use different values for the objective function, but they all are based on flow equilibrium conditions that can be made to cope with dynamic traffic flows in practical cases.

This research work is essentially intended to give a contribution to the research field of freight distribution within mixed-traffic, as well as to the distribution system of a PSO-SOC within the proposed assumptions.

REFERENCES

1. S. Soehodho, Nahry, Strategic design of distribution system of state-owned companies: preliminary stage of logistics research series and evaluation on model parameters, International Journal on Logistics and Transportation (IJLT) 3 (1) (2009).
2. K.S. Bhutta, F. Huq, G. Frazier, Z. Mohamed, An integrated location, production, distribution and investment model for a multinational corporation, International Journal Production Economics 86 (2003) 201–216.
3. M. Sun, Solving the uncapacitated facility location problem using tabu search, Computers and Operations Research 33 (2006) 2563–2589.
4. L. Dupont, Branch and bound algorithm for a facility location problem with concave site dependent costs, International Journal Production Economics 112 (2008) 245–254.
5. E.S. Savas, On equity in providing public services, Management Science 24 (8) (1978) 800.
6. G.T. Ross, R.M. Soland, A multicriteria approach to the location of public facilities, European Journal of Operational Research 4 (1980) 307–321.
7. S. Yan, D.S. Juang, C.R. Chen, W.S. Lai, Global and local search algorithms for concave cost transshipment problems, Journal of Global Optimization 33 (2005) 123–156.
8. J. Harkness, C. ReVelle, Facility location with increasing production costs, European Journal of Operational Research 145 (2003) 1–13.

9. E.T. Iakovou, An interactive multiobjective model for the strategic maritime transportation of petroleum products: risk analysis and routing, Safety Science 39 (2001) 19–29.
10. K. White, L.R. Grenzeback, Understanding freight bottlenecks, Public Roads 70 (5) (2007).
11. C. Jones, Perspective on freight congestion, Public Road, ProQuest Science Journals 36 (2007).
12. F. Glover, D. Klingman, N.V. Phillips, Network Models in Optimization and Their Applications in Practice, John Wiley & Sons, Inc., 1992
13. R.K. Ahuja, T.L. Magnanti, J.B. Orlin, Network Flows, Prentice Hall, New Jersey, 1993.

CITATION

Sutanto Soehodho, Nahry, Traffic flow consideration in design of freight distribution system, IATSS Research, Volume 34, Issue 1, July 2010, Pages 55-61, ISSN 0386-1112, http://dx.doi.org/10.1016/j.iatssr.2010.06.007.

CHAPTER 8

Evaluation of Urban Redevelopment Impact on Non-Motorized Traffic

Yanli Wang[1], Liang Li[2], Xiaoyu Zhu[3], Bing Wu[1], Linbo Li[1]

[1] Key Laboratory of Road and Traffic Engineering of Ministry of Education, Tongji University, Shanghai 201804, China

[2] China Sinogy Electric Engineering Co., Ltd., Shanghai 200061, China

[3] National Center for Smart Growth Research and Education, University of Maryland, College Park, MD 20742, USA

ABSTRACT

As an important component of city evolution, urban land redevelopment has an impact on transportation system. The current traffic impact analysis (TIA) is lack of a comprehensive component for non-motorized transportation under redevelopment. For a better guidance of land redevelopment and non-motorized transportation planning, it is necessary to evaluate the negative impact of redevelopment on non-motorized traffic in the TIA. In this paper, an evaluation framework for the impact analysis is built up. We organized the procedures and components of impact evaluation, and proposed the corresponding qualitative and quantitative evaluation indicators for non-motorized traffic under redevelopment. Level of service (LOS) and its criterion are employed for external impact evaluation, and level of safety, convenience, independence, and comfort which are four aspects of quality of service (QOS) are proposed to analyze the internal impact. The framework is applied to a redevelopment study in Shanghai, China. The case study results indicate that the redevelopment from a residential area to a mixed commercial area has a significant impact on non-motorized traffic. The potential negative impact from both external and internal traffic can be minimized by reasonable improvements in the internal land use design.

INTRODUCTION

Currently, with the rapid development of urbanization, land redevelopment becomes an importation part of city evolution in addition to urban sprawl and new town construction. This is especially important for the metropolitans where the available building land is limited and the part of the pre-existing land use is not efficient to afford the increasing development in population and economy. Therefore, land redevelopment is critical to maintain the urban vitality and cities upgrade (Friedman et al., 2004 and Woodbury and Bauer, 1953). Many researchers from urban planning and economy attempted to utilize the urban area fully by studying when, where and how to redevelop (Capozza and Sick, 1991, Childs et al., 1996 and Williams, 1997). However, redevelopment causes a lot of issues on travel behavior, transportation system, and travel demand, even though it improves the efficiency of land use.

Unlike the new development, urban redevelopment is a reconstruction on a previously developed area. Specifically, redevelopment changes the land use between types of urban land, while new development is changing non-urban land into urban land. Usually, development locates in suburbs, and redevelopments locate in old urban areas with more complex surrounding environment and land use. Therefore, compared to building new facilities in new development projects, redevelopment projects also need to deal with the internal and external existing transportation facilities and travel demand especially non-motorized traffic (Koll-Schretzenmayr, 1999).

On one hand, non-motorized travel including walking, bicycling and bicycle-derived modes such as scooters (Guitink and Flora, 1995, Iacono et al., 2010, Plaut, 2005 and Rietveld, 2001) is the main travel mode in old cities. For example, non-motorized mode share is more than 50% in Shanghai, China (Xiong et al., 2010). It is a healthy and sustainable travel mode (Rietveld, 2001), and plays an essential role in urban transport in most of the developing countries (Replogle, 1992). However, facilities and environment for the non-motorized traffic are required but are often ignored in transportation planning, and the impact on the traffic is not well evaluated. On the other hand, when land use changes under redevelopment, the travel demand and behavior including non-motorized travel will be influenced (Guo et al., 2007 and Rodríguez and Joo, 2004), and it is necessary to evaluate the impact of redevelopment in the project proposal. Therefore, many previous studies formed a relatively complete transportation impact analysis system in United States and other countries (Dey and Fricker, 1994 and Muldoon and Bloomberg, 2008). However, only a few of them emphasized the impact of redevelopment (Hulse et al., 2004 and Wang et al., 2011), especially for the non-motorized traffic.

This study focuses on constructing an evaluation framework and criterions of non-motorized traffic under urban redevelopment. This framework can be applied to analyze the impact of land redevelopment and provide suggestions for decision making and land use design. The impact mentioned in this paper refers to negative impact. The objective of this study is to fulfill the gap of land use redevelopment and non-motorized traffic evaluation in traffic impact analysis (TIA), which is critical to support the redevelopment plan and minimize the potential negative impact. In the next section, the impact mechanism and factors of land redevelopment on the non-motorized traffic are summarized. After that, the framework with evaluation procedures and corresponding indicators for non-motorized traffic are proposed and described in detail in Section 3. In Section 4, a case study of Shanghai, China is analyzed as an example of this framework and the results are discussed. The important conclusions and further research are emphasized in the final section.

IMPACT MECHANISM ON NON-MOTORIZED TRAFFIC DEMAND

In this section, we focus on land use, facilities and environments that may be changed in redevelopment and their impact on non-motorized traffic. Land is the source of social activities, and decides the generation of activities and the trips temporally and spatially. In addition, the traffic system carries the traffic with limited capacities. Therefore, traffic demand is subject to the dual constraints of land use and transportation facilities.

Land use factors

Non-motorized traffic is highly affected by land use factors. For example, compact, mixed use neighborhoods have more non-motorized shopping or recreational travel than auto-oriented neighborhoods (Cervero and Radisch, 1996 and Handy, 1996). Required by city expansion and profit maximization, redevelopment is usually accompanied with transformation in function (such as the changing of the residential land for commercial purposes) and increase in land use intensity. Therefore, the results of the trip generation, trip distribution, mode choice and the final trip assignment will be various from previous redevelopment.

First, when the function of the land use changes, the trip purpose, rate and mode share will be different. For example, according to "Technical Standards of Traffic Impact Analysis of Shanghai Construction Project" (SUPDRI, 2010), if a plot of residential land with shabby houses is redeveloped into commercial office building in Shanghai downtown, the daily trips rate will be 0.30 per construction square instead of 0.07. And the non-motorized share will be 32% after redevelopment instead of 50% before redevelopment. Also the trip generation by time of day is different, e.g. the trip production coefficient of peak hour is 0.20 before, but about 0.05 after; and the trip attraction coefficient of peak hour is not more than 0.05 before, but about 0.25 after.

Second, the intensity of land use is also correlated with the trip density. In the above example, if the site with shabby houses is rebuilt into high-rise housing, then the daily trip rate will be still around 0.07 per construction square, but the total number of trips will increase, and the non-motorized mode share will decline by nearly 10%.

Overall, both the function and intensity of the redevelopment site will affect the non-motorized traffic demand. Then combined with surrounding land use functions, the short distance trips will redistribute and the final non-motorized traffic will also reallocate spatially. For example, a residential site surrounded by housing units, is redeveloped into a shopping center. Then it will attract most of the shopping trips by surrounding residents from other shopping destinations. Those home-based shopping trips are short and more likely to be non-motorized. On the contrary, if a shopping center surrounded by housing is redeveloped into housing, the results are opposite.

Transport facilities and traffic environment

Transport facilities and traffic environment after the redevelopment are also correlated to non-motorized share, route and safety (Cao et al., 2009 and Reynolds et al., 2009). An appropriate design can increase quality and quantity of non-motorized traffic, such as provision of crossing facilities, exclusive bicycle lanes, as well as a comfortable sidewalk (Mehta, 2008, Saelens et al., 2003, Shriver, 1997 and Sisiopiku and Akin, 2003). In addition, the coordination of the redevelopment with surrounding transportation facilities will also impact the non-motorized traffic. For example, the connection of the site to the external traffic system with many conflict points or longer detour distance will decrease non-motorized mode share because of the safety issue and inconvenience.

RESULTS AND DISCUSSION

Evaluation process

Similar to general TIA, data collection, traffic demand forecast, and evaluation are also the main steps of the redevelopment impact evaluation on non-motorized traffic. But the process and contents are not the same with general TIA due to the uniqueness of redevelopment and non-motorized traffic. The detailed steps are shown in Fig. 1. (1) Data collection stage including Steps 1, 3, 4 in Fig. 1, needs focus on non-motorized traffic and facilities: non-motorized traffic (pedestrian and bicyclists) volumes, road crossing facilities, bicycle parking, sidewalk, path and trail conditions (effective width, surface condition, sight distances, etc.), special hazards to walking and bicycling, security, cleanliness, vandalism, aesthetic conditions, and so on. (2) Traffic demand forecast stage including Steps 5, 6 in Fig. 1, needs considering the original traffic demand and facilities in the redeveloped site before redevelopment and the surrounding traffic demand transfer due to redevelopment. (3) Evaluation stage comprised of Steps 2, 7, 8, 9, 10 in Fig. 1, needs different indicators to do the analysis, which is in detail described in Section 3.2. (4) Transportation management and improvement including Steps 11, 12 in Fig. 1 are also important, and detail measures are given in Section 3.3.

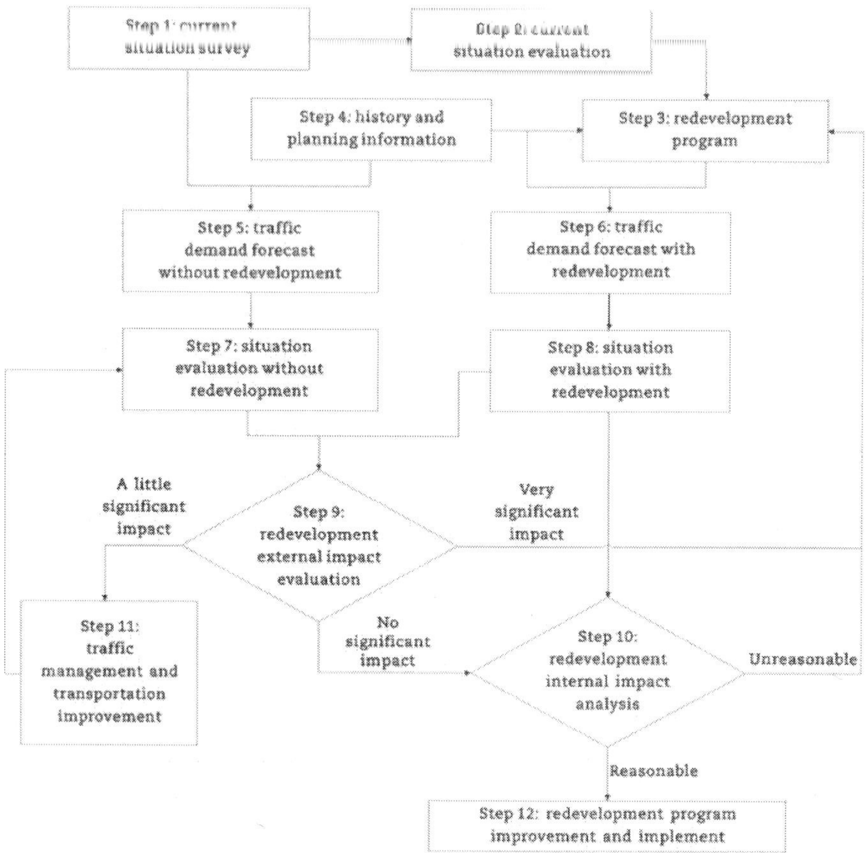

Figure 1. Traffic impact evaluation process of redevelopment.

Impact analysis fields and evaluation indicators

Fig. 2 shows the relationship between internal and external traffic of redevelopment. Here, internal refers to the redevelopment site itself, and external is the outside or surrounding of the site. A synthesized TIA should integrate impacts on motorized and non-motorized traffic including solid and broken lines in Fig. 2. Here we focus on the impact evaluation on non-motorized traffic covered by the solid lines in Fig. 2. Including the impact on external and internal non-motorized traffic, it has different objects and evaluation indicators from the general TIA of motorized traffic.

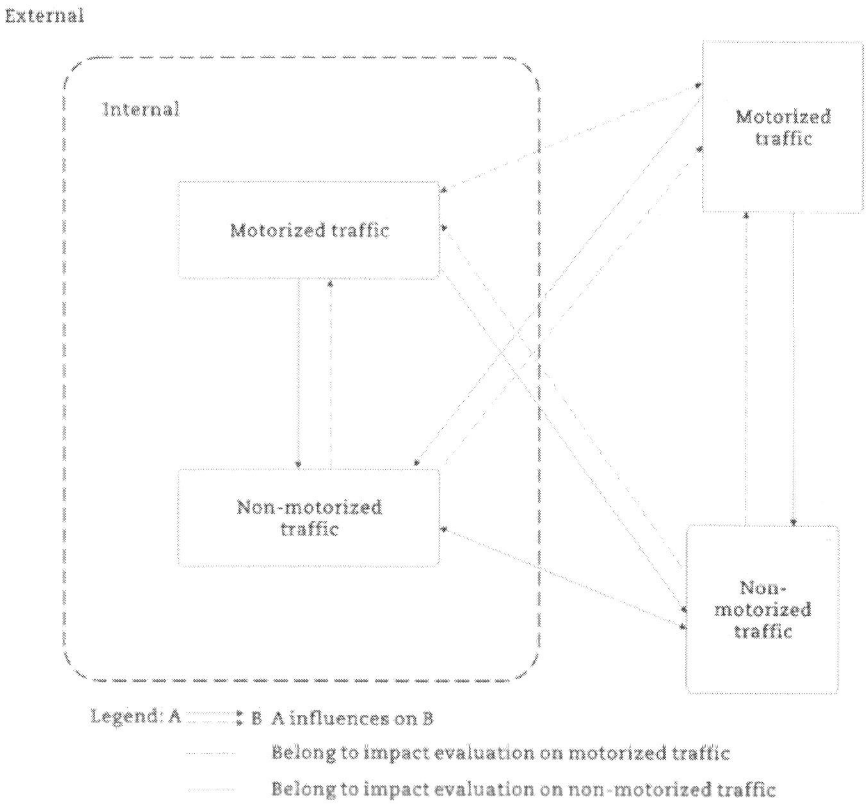

Figure 2. Traffic impact aspects of redevelopment.

Impact analysis on external non-motorized traffic

The impacts on external non-motorized traffic come from not only the external non-motorized traffic demand changes but also the external motorized traffic demand changes due to redevelopment. And as shown in Fig. 2, the transform of internal motorized and non-motorized traffic demand also affects the external non-motorized traffic through internal and external connection such as entrances and exits. The impact factors include the changes of land use, facilities and environment.

The prerequisites to evaluate the impact on external non-motorized traffic are definition and criterion of the evaluation indicators on two external objects: roadway section and crossing. As all know, level of service (LOS) has been used to evaluate motorized transportation facilities for a long time, and since Botma (1995) proposed LOS methodologies for bicycle paths and pedestrian paths, the LOS criteria for exclusive and shared paths are adopted by Highway Capacity Manual (TRB, 2000). Various LOS evaluation methods and criteria are also studied in the last few years (Dowling et al., 2008, Patten et al., 2006 and Phillips et al., 2001). In fact, the evaluation methods and criteria are not unique depending on the countries and regions. Currently, there are few studies on evaluation methods and criteria for bicycle and pedestrian paths in China where non-motorized traffic volume are huge. Here, we introduce level of service (LOS) to evaluate the external non-motorized traffic situation. Same as HCM, the LOS ratings of transportation facilities range from A (best) to F (worst, or failure).

In this paper, considering the data availability and comparability, we use *V/C* ratio (*V* is hourly traffic volume, and *C* is capacity) to determine LOS. And considering the crowded situation of non-motorized traffic in China, we provide an evaluation criterion for LOS of bicycle and pedestrian lanes in China urban area in Table 1 based on the observation and manual (TRB, 2000). And the capacity of the lanes can be calculated according to national manual (MOC, 1995) and the local manuals.

Table 1. Evaluation criterion for LOS of non-motorized paths in urban area in China.

LOS	A	B	C	D	E	F
V/C ratio	≤0.20	0.20–0.35	0.35–0.55	0.55–0.70	0.70–0.85	>0.85

In addition, to judge the impact due to redevelopment, all kinds of non-motorized traffic facilities (including the bicycle lanes, sidewalks and the crossing facilities) need to be evaluated to obtain the LOS. And then we compare the LOS with and without redevelopment to evaluate the impact. To some extent, promoting non-motorized traffic is good for sustainable development; however increasing a large number of non-motorized traffic may reduce LOS a lot and cause significant negative impact on external traffic. For example, it is good to change the LOS from A to B or C, but bad to change LOS from E to F. According to our survey of Shanghai residents and experts about the non-motorized traffic facilities service in 2008, a significant negative impact evaluation criterion for non-motorized traffic is suggested to determine the impact degree (Table 2), which is consistent with the motorized traffic impact evaluation criterion (MOHURD, 2010). For example, if the LOS without redevelopment is B, then the impact is significant negative when the LOS with redevelopment is D or E, and it is insignificant negative or positive when the LOS with redevelopment is C or B.

Table 2. Significant impact evaluation criterion on non-motorized traffic in urban area in China.

LOS without redevelopment	A	B	C	D	E	F
LOS with redevelopment	D or higher	D or higher	D or higher	E or higher	F or higher	F or higher

Impact analysis on internal non-motorized traffic

The impacts on internal non-motorized traffic are mainly caused by the changes in internal motorized and non-motorized traffic demand after redevelopment. Besides, the external traffic can also influence the internal traffic through internal and external connections (Fig. 2). The impact factors are mainly land use and facilities of the site itself. So the negative impacts can be easily reduced through reasonable design of the redevelopment project.

Evaluating the impact on internal non-motorized traffic is very complicated because the facilities are redesigned under redevelopment. The measurement of the impacts can be achieved by analyzing the non-motorized traffic performance of the whole site or community instead of comparing LOS of specific facilities. When evaluating a community, quality of service (QOS) is usually used in United States as the overall measurement or perceived performance of service from the passenger's or user's point of view (Phillips and Guttenplan, 2003). In this paper, specific indicators are used instead of a comprehensive indicator of QOS so as to improve the inadequacies, because our target is to decrease the negative impacts of redevelopment on non-motorized traffic. Therefore, from the user's perspective, the performance of QOS is divided into four aspects of service: safety, convenience, independence, and comfort. The performance measures could be quantitative or qualitative factors.

The rationality of community design can be analyzed using the above four indicators. Each aspect of service reflects the overall community performance composed of dynamic (such as traffic flow on the roads) and static (such as parking service) systems. The measurement of each indicator is shown in Fig. 3. (1) The conflict point is employed to analyze safety of the internal and connection design. (2) Detour time is selected to reflect the convenience of the layout, such as the detour time to the transit stations or taxi stop. (3) The independence level is measured by the mix level of motorized and non-motorized traffic at roadways and parking lots. (4) Comfort means the community is environmentally friendly with high greening rate and humanized design.

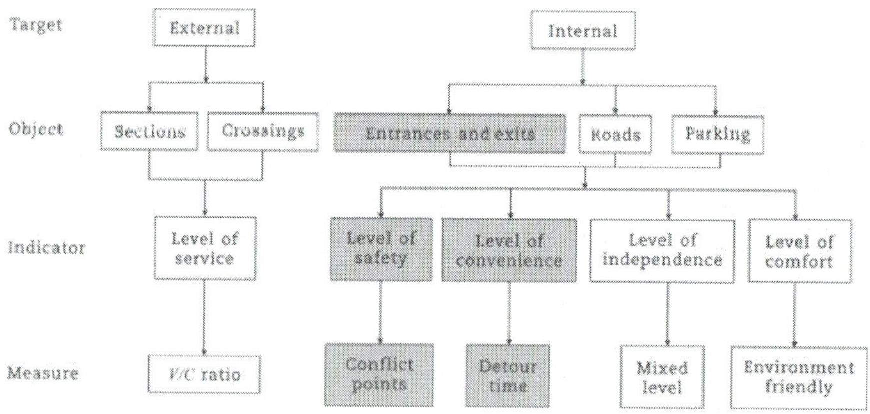

Figure 3. Impact evaluation indicators and measures of non-motorized traffic.

In addition, before evaluating impact on internal non-motorized traffic, the community design (such as lane width, signs, location and number of entrance) must comply with the relevant design standards. Therefore, the roadways inside the community are designed according to the traffic demand after redevelopment, and we don't analyze the level of service of the traffic inside.

Finally, based on the above discussion, all impact evaluation indicators and the related measures of non-motorized traffic are displayed in Fig. 3. This indicator system is easy to understand and apply, and can also be extended to the general non-motorized impact analysis.

Transportation management and improvement

At the end of the process (Fig. 1), the external impact level or degree is clearly demonstrated after Step 9. If the negative impact is huge, the redevelopment project scenarios including the land use property and intensity (Step 3) need to be adjusted, until the impact evaluation is appropriate. If the impact is a little significant, the transportation management and improvement (Step 11) are necessary to improve the design. If the impact is moderate, but the result of the internal impact analysis (Step 10) is unreasonable, then it is required to adjust the internal transportation of the redevelopment project (Step 3). This section focuses on adjusting the redevelopment project design or improving the external facilities to decrease the negative impact due to redevelopment.

The internal non-motorized traffic organization and management should be optimized before they are implemented. The target is to maintain the traffic safety within the community through reducing the conflicts between non-motorized and motorized traffic. Independent non-motorized facilities could be designed to keep pedestrian and bicyclist safe and comfort. And the design should consider the accessibility and convenience of non-motorized traffic. Meanwhile, the connection of the site to the external traffic system is crucial to reduce the impact on non-motorized traffic, and the negative impact can be adjusted by relocating the entrance or exit. For example, one or two entrances (or exits) exclusive for non-motorized traffic can improve the convenience for residents or workers by decreasing detour time so as to encourage people use non-motorized mode. And if the entrance for non-motorized traffic is close to bus stops or crossing facilities, detour time can also be reduced. In addition, if the main entrance (or exit) for motor vehicles locates on the street with fewer bicycles or pedestrians, the impact on the external non-motorized traffic will be reduced. The motor vehicles can be forced to right in and out at the entrances and exits to minimize the conflict points. Another effective method is forbidding motorized vehicles to occupy the non-motorized facilities such as the lanes near the connection.

The external traffic management and non-motorized facilities improvement are mostly ignored and also difficult to achieve. The main aspects include: (1) Redesign the crossing facilities on road sections or the intersections that are highly influenced by redevelopment. For example, in the segments where the non-motorized entrance located, adding (or changing) the crossing facilities or installing the central fences to avoid jaywalking; or broadening the street crosswalks at the intersections if there are more people crossing the intersection after redevelopment. (2) Redesign the traffic signal control at the main affected intersections, such as adding a non-motor vehicle phase or adjusting the pedestrian green length and cycle length. The adjustment of the signal control requires a specific calculation according to the predicted motorized and non-motorized traffic at the intersection after redevelopment. (3) Redesign the critical road segments and intersections, such as widening non-motorized vehicles lanes or sidewalks, or providing left turn bay for bicycles at the intersection.

These methods can mitigate the negative redevelopment impact to a certain extent. However, if the impact is huge or still beyond an acceptable level through a variety of improvements, the redevelopment project must be fundamentally redesigned such as revise the property and intensity of land redevelopment.

CASE STUDY

A redevelopment site of Shanghai is chosen as a case study to analyze the impact of redevelopment on non-motorized traffic using the framework we proposed. The background and the results are discussed in this section.

Background

Shanghai is the largest city of population and a major financial center in China, located in the Yangtze River Delta. By the end of 2008, the construction land area of Shanghai is 2288 km^2 with a population of 20 million, while the registered number of bicycle and electronic bicycle is 4.05 million and 2.77 million respectively, and 54.9 percent of trips are made by non-motorized modes according to "the fourth comprehensive traffic survey report of Shanghai" (SHUCM, 2010).

The urban area of Shanghai is composed of eight districts. The study site is located at the junction of three districts (Changning, Jing'an, and Putuo). It is an important part of Caojiadu Community, Changning District, surrounded mainly by residential area. This redevelopment is initialed in 2006 and constructed from the end of 2008 to the end of 2012, and this impact evaluation study is conducted in early 2008. Fig. 4 shows the location and the surrounding land use of the study site.

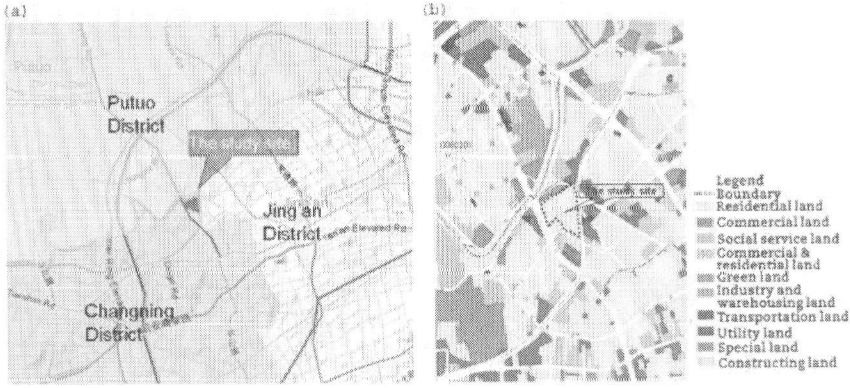

Figure 4. Location and surrounding land use of study site. (a) Location of study site. (b) Surrounding land use.

The site area is about 46,703 m^2, and it was an exclusive residential land before redevelopment. The site is planned to redevelop into a mixed land use of residential and commercial utilities. It includes 97,622 m^2 building area of housing units, 51,968 m^2building area of offices, and 40,421 m^2 building area of shopping center. The land use and roadway network around the study site before and after redevelopment and their changes are displayed in Fig. 5. Road Section a-b is arterial road with separated sidewalks on both sides, and crosswalks are only existing in Intersections a and b.

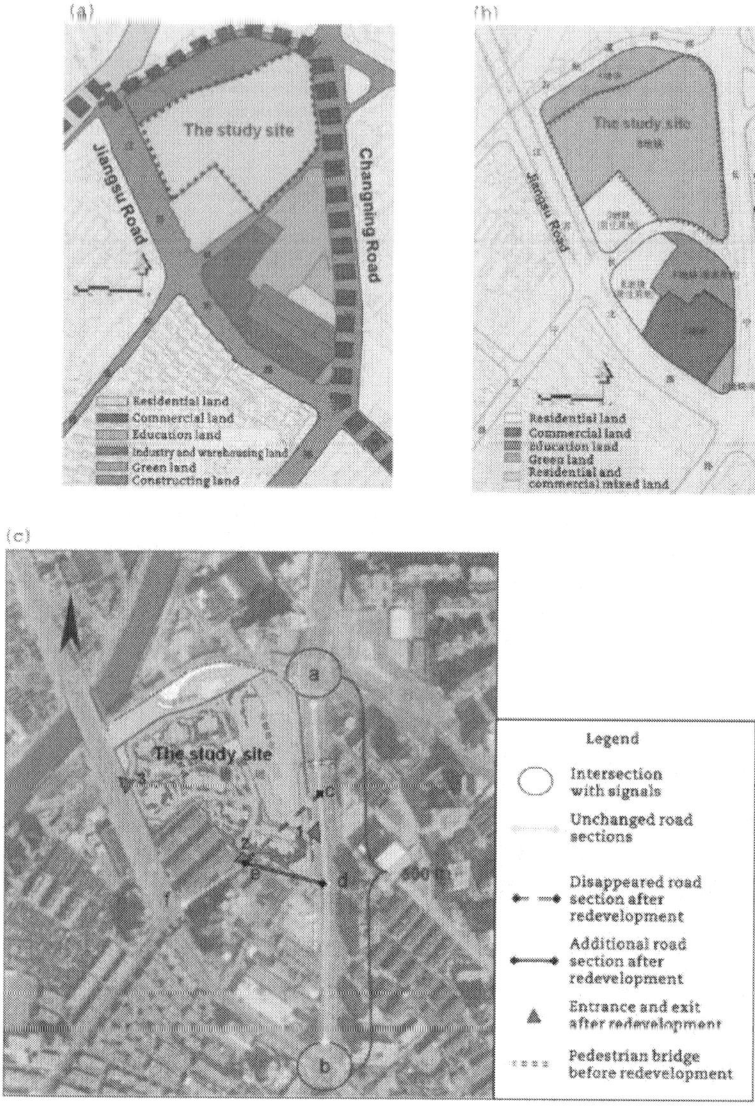

Figure 5. Land use and road network around study site before and after redevelopment. (a) Before redevelopment. (b) After redevelopment. (c) External facilities changes.

The transportation data (including facilities and traffic information) are collected from our survey in 2008. We investigated the physical characteristics of the facilities, such as design of roadway section and signal control of intersections from June 3rd to 9th, 2008. And we did the survey of operating conditions of the facilities such as the volume of motorized vehicles, bicycles and pedestrians in peak and off-peak hours of weekday in the following week. In addition, the history and planning data were obtained from the Bureau of Planning, Land and Resources of Shanghai and regional planning and land authority. The design scenario came from the real estate developer. The trip rate and mode choice information were adjusted according to the survey combining with "the third comprehensive survey report of Shanghai" (Lu et al., 2005). We chose the extent of general TIA as the study area and a range of one kilometer outside the site boundary as the main impact area. The peak hour 4 pm–6 pm with more non-motorized traffic is selected as evaluated period. Meanwhile, according to TIA (MOHURD, 2010), the impact evaluation year is 2015 which is three years after the project finished.

After data collection, we obtained the necessary information for forecasting traffic demand of all travel modes with and without redevelopment for 2015. The traffic demand without redevelopment is based on current traffic condition and the conventional regional and local development. The traffic demand with redevelopment depends on the property and density of the redeveloped site such as how many residents, jobs and visitors it can attract. And the traffic forecast should include both motorized traffic and non-motorized traffic because the non-motorized traffic is also influenced by motorized traffic. In addition, the difference also exists in transportation facilities with and without redevelopment. Therefore, the land use characteristics (such as households and jobs) and transportation facilities with and without redevelopment are different for traffic demand models. Here we only show the results because the forecast process is complex and not the focus.

Results

Based on our survey and forecast, the impact analysis and evaluation are conducted for the study site, focusing on the impact on non-motorized traffic. The results are discussed for external and internal separately in this section. The target is to reduce the negative impact to acceptable level. If the impact to external traffic cannot be acceptable with design improvement, it is necessary to modify the property or density of the redevelopment project. Also, the impact evaluation on the internal non-motorized traffic and the improvement of traffic system should meet the requirement in safety, convenience, independence and comfort.

The impact on external traffic

The roadway segments between Intersection a and Intersection b are strongly affected. The *V/C* ratios of these segments with and without redevelopment are provided in Table 3. And Table 3 also shows the LOS of the facilities according to Table 1.

Table 3. Service level changes of main impact links.

		Southbound		Northbound	
		Sidewalk	Bicycle lane	Sidewalk	Bicycle lane
Section a-c *V/C* (LOS)	Without redevelopment	0.41(C)	0.43(C)	0.39(C)	0.45(C)
	With redevelopment	0.58(D)	0.52(C)	0.55(C)	0.50(C)
Section c-d *V/C* (LOS)	Without redevelopment	0.31(B)	0.40(C)	0.42(C)	0.44(C)
	With redevelopment	0.41(C)	0.53(C)	0.54(C)	0.49(C)
Section d-b *V/C* (LOS)	Without redevelopment	0.32(B)	0.40(C)	0.37(C)	0.43(C)
	With redevelopment	0.40(C)	0.54(C)	0.44(C)	0.48(C)

Combined with Table 2, Table 3 implies the impact on southbound sidewalk of Section a-c is significant. The impact can be reduced by increasing the width of the sidewalk. This is easy to be accomplished because this side is along the designed shopping street to be constructed and there is enough space to increase the width of the sidewalk. With a wide sidewalk, the *V/C* is less than 0.4 (LOS is C), and the negative significant impact is reduced to the acceptable level.

In addition, the LOS of non-motorized traffic at the intersections will not change significantly with redevelopment, but the signal timing (such as cycle length and the pedestrian green length) still needs to be adjusted according to the traffic demand after redevelopment.

Furthermore, in this case, some special impacts occur due to the road Section e-c changing to e-d (Fig. 5(c)), which cannot be evaluated by LOS. This change can influence the detour distance and path selection of non-motorized traffic. Actually, it brings negative impact to the residents and employees in the south of road Section e-f, which is only a small portion. Meanwhile, the detour distance to the north will only increase about

100 m, which is 90 s by walk, and 30 s by bike. Therefore this impact is acceptable.

Besides, it is not reasonable that there is no crossing facility between Section c-b before redevelopment. This shortage can be overcome by designing a crosswalk with signal control or pedestrian bridge at the location of Point d in Fig. 5(c) after redevelopment. And this signal control needs to be optimized together with the signal control of Intersections a and b.

The impact on internal traffic

Once the design of the redevelopment project meets the requirement of external traffic system, the internal traffic system of the redevelopment project itself is analyzed to make sure the rationality of the design. Fig. 6 displays the layout of internal buildings and traffic facilities, and the motorized and non-motorized traffic flow lines.

Figure 6. Internal design of motorized and non-motorized traffic flow lines.

Here we evaluate the impact and reasonableness of the redevelopment project according to the indicators of internal traffic which are proposed in Section 3.2. (1) It is safer after redevelopment, because the site will have independent non-motorized traffic roads to replace the previous mixed lanes, and the conflict points only exist in the motorized entrances and exits. (2) It is more convenient after redevelopment, because in order to reduce the detour distance and enhance convenience, many improvements are considered such as designing the non-motorized traffic roads in the middle of the site, and adding some non-motorized entrances and exits near the transit and taxi stops. (3) Bicycle riders have their own parking lot and independent road system, so the level of independence is higher. (4) More greening and public facilities are provided with a dedicated property management after redevelopment, and the commercial area is separated from the residential areas in order to reduce noise and disturbance to residents. Therefore, it is more comfortable and livable than before. In general, the internal design of the redevelopment is reasonable.

Summary

From the above analysis, we obtain that the redevelopment project can bring significant impact to external non-motorized traffic, but this impact can be reduced by adjusting the internal design. After improvements, the impact on non-motorized traffic is acceptable. In addition, the relevant agencies need to monitor the effective implementation of the improvements.

Besides, in this case, redeveloping residential land into mixed of shopping center, office building and housing has a significant increase in the pedestrian flow on the sidewalk near the site, and a wider sidewalk is suggested compared to the original plan. The impact on bicycle traffic is not significant, and the non-motorized lanes can be broadened later combining with the improvement plan of this road segment. Also the impact caused by facility changing is acceptable in this study.

CONCLUSIONS

Currently, land redevelopment becomes an important component of city evolution, especially in metropolitans with the constraint in land resource. The redevelopment based on existing urban land use and facilities can change either land use function or intensity, or both. This paper emphasizes the importance of the urban redevelopment impact evaluation on non-motorized traffic, and proposes a framework of redevelopment impact evaluation to fulfill the gap in the current TIA. In the framework, the impact evaluation indicators and measures of non-motorized traffic are discussed. Level of service (LOS) represented by *V/C* ratio is used as the evaluation criterion of external bicycle and pedestrian traffic. The performance of quality of service (QOS) including safety, convenience, independence, and comfort is proposed to evaluate the internal non-motorized traffic.

In this paper, we also utilize a real redevelopment project as an example to explain the framework and each component. As a result of the case study, there are significant impacts on non-motorized traffic when a residential land redeveloped into a mixed commercial land use. It is necessary to conduct the impact evaluation on non-motorized traffic for the redevelopment transportation planning and minimize the negative impact. Based on the evaluation results, reasonable suggestions for the design modification with minimum cost are provided to reduce the negative impact. For example, increasing the width of the sidewalk at the side of redevelopment site is required to accommodate the increasing pedestrian traffic, and designing a crosswalk with signal control close to the bus stop will improve the connectivity. The safety, convenience, independence and comfort requirements for non-motorized traffic are achieved by the proper internal design, such as divided bike lanes and parking for motorized and non-motorized traffic, and non-motorized entrances and exits near the transit and taxi stops. In addition, the impacts caused by facility change need attentions as well. Overall, in this case the negative impact can be reduced by adjusting the internal design.

There are necessity and potential for substantial improvement in the context of this framework. In the future, more case studies of redevelopment are planned to be investigated to generate a more general evaluation criterion. The framework and process can be integrated to the transportation impact analysis system.

ACKNOWLEDGMENTS

This research was supported by the Specialized Research Fund for the Doctoral Program of Higher Education (No. 200802470030).

REFERENCES

1. Botma, H., 1995. Method to determine level of service for bicycle paths and pedestrian-bicycle paths. Transportation Research Record 1502, 38e44.
2. Cao, X.Y., Mokhtarian, P.L., Handy, S.L., 2009. The relationship between the built environment and nonwork travel: a case study of Northern California. Transportation Research Part A: Policy and Practice 43 (5), 548e559.
3. Capozza, D.R., Sick, G.A., 1991. Valuing long-term leases: the option to redevelop. The Journal of Real Estate Finance and Economics 4 (2), 209e223.
4. Cervero, R., Radisch, C., 1996. Travel choices in pedestrian versus automobile oriented neighborhoods. Transport Policy 3 (3), 127e141.
5. Childs, P.D., Riddiough, T.J., Triantis, A.J., 1996. Mixed uses and the redevelopment option. Real Estate Economics 24 (3), 317e339.
6. Dey, S.S., Fricker, J.D., 1994. Traffic impact analysis and impact fees in state departments of transportation. ITE Journal 64 (5), 39e44.
7. Dowling, R., Flannery, A., Landis, B., et al., 2008. Multimodal level of service for urban streets. Transportation Research Record 2071, 1e7.
8. Friedman, M.T., Andrews, D.L., Silk, M.L., 2004. Sport and the fac¸ade of redevelopment in the postindustrial city. Sociology of Sport Journal 21 (2), 119e139.
9. Guitink, P., Flora, J., 1995. Non motorized transportation in transportation systems: back to the future. In: Transportation Research Board 74th Conference, Washington DC, 1995.
10. Guo, J.Y., Bhat, C.R., Copperman, R.B., 2007. Effect of the built environment on motorized and nonmotorized trip making: substitutive, complementary, or synergistic? Transportation Research Record 2010, 1e11.
11. Handy, S.L., 1996. Urban form and pedestrian choices: study of Austin neighborhoods. Transportation Research Record 1552, 135e144.
12. Hulse, K., Herbert, T., Down, K., 2004. Kensington estate Redevelopment Social Impact Study. The Swinburne Institute for Social Research, Swinburne University of Technology, Melbourne.
13. Iacono, M., Krizek, K.J., El-Geneidy, A., 2010. Measuring nonmotorized accessibility: issues, alternatives, and execution. Journal of Transport Geography 18 (1), 133e140.
14. Koll-Schretzenmayr, M., 1999. From greenfield development to brownfield redevelopment. disP-The planning Review 35 (139), 43e48.
15. Lu, X., Xue, M., Zhu, H., 2005. The Third Comprehensive Survey Report of Shanghai. Shanghai City Comprehensive Transportation Planning Institute, Shanghai.
16. Mehta, V., 2008. Walkable streets: pedestrian behavior, perceptions and attitudes. Journal of Urbanism 1 (3), 217e245.

17. MOC, 1995. Code for Transport Planning on Urban Road. Ministry of Construction of the People's Republic of China, Beijing. GB50220-95.
18. MOHURD, 2010. Technical Standards of Traffic Impact Analysis of Construction Projects. Ministry of Housing and Urban-Rural Development of the People's Republic of China, Beijing.
19. Muldoon, D., Bloomberg, L., 2008. Development of best practices for traffic impact studies. Transportation Research Record 2077, 32e38.
20. Patten, R.S., Schneider, R.J., Toole, J.L., et al., 2006. Shared-use Path Level of Service Calculator-a User's Guide. Federal Highway Administration, Mclean.
21. Phillips, R.G., Guttenplan, M., 2003. A review of approaches for assessing multimodal quality of service. Journal of Public Transportation 6 (4), 69e87.
22. Phillips, R., Karachepone, J., Landis, B., 2001. Multi-modal Quality of Service Project. Project report for. Florida Department of Transportation, Tallahassee.
23. Plaut, P.O., 2005. Non-motorized commuting in the US. Transportation Research Part D: Transport and Environment 10 (5), 347e356.
24. Replogle, M., 1992. Non-motorized Vehicles in Asian Cities. National Technical Information Service, Alexandria.
25. Reynolds, C.C.O., Harris, M.A., Teschke, K., et al., 2009. The impact of transportation infrastructure on bicycling injuries and crashes: a review of the literature. Environmental Health 8, 1e19.
26. Rietveld, P., 2001. Biking and Walking: the Position of Nonmotorized Transport Modes in Transport Systems. Vrige University, Amsterdam.
27. Rodrı´guez, D.A., Joo, J., 2004. The relationship between nonmotorized mode choice and the local physical environment. Transportation Research Part D: Transport and Environment 9 (2), 151e173.
28. Saelens, B.E., Sallis, J.F., Frank, L.D., 2003. Environmental correlates of walking and cycling: findings from the transportation, urban design, and planning literatures. Annals of Behavioral Medicine 25 (2), 80e91.
29. Shriver, K., 1997. Influence of environmental design on pedestrian travel behavior in four Austin neighborhoods. Transportation Research Record 1578, 64e75.
30. SHUCM, 2010. The Fourth Comprehensive Traffic Survey Report of Shanghai. Shanghai Urban Construction and Communications Commission, Shanghai.
31. Sisiopiku, V.P., Akin, D., 2003. Pedestrian behaviors at and perceptions towards various pedestrian facilities: an examination based on observation and survey data. Transportation Research Part F: Traffic Psychology and Behaviour 6 (4), 249e274.

32. SUPDRI, 2010. Technical Standards of Traffic Impact Analysis of Shanghai Construction Project (Draft). Shanghai Urban Planning and Design Research Institute, Shanghai.
33. TRB, 2000. Highway Capacity Manual. Transportation Research Board, Washington DC.
34. Wang, Y., Li, L., Wu, B., 2011. The mechanism of urban land redevelopment impact on traffic systems. In: Proceedings of 11th ICCTP, Nanjing, 2011.
35. Williams, J.T., 1997. Redevelopment of real assets. Real Estate Economics 25 (3), 387e407.
36. Woodbury, C., Bauer, C., 1953. The Future of Cities and Urban Redevelopment.
37. University of Chicago Press, Chicago. Xiong, W., Chen, X., Hu, X., 2010. Discussion on planning for nonmotorized travel in cities. Urban Transport of China 8 (1), 44e54.

CITATION

Yanli Wang, Liang Li, Xiaoyu Zhu, Bing Wu, Linbo Li, Evaluation of urban redevelopment impact on non-motorized traffic, Journal of Traffic and Transportation Engineering (English Edition), Volume 2, Issue 3, June 2015, Pages 187-197, ISSN 2095-7564, http://dx.doi.org/10.1016/j.jtte.2015.03.007.

CHAPTER 9

A Simulated Environment Experiment on Annoyance Due to Combined Road Traffic and Industrial Noises

Catherine Marquis-Favre and Julien Morel

Univ Lyon, Ecole Nationale des Travaux Publics de l'Etat, Laboratoire Génie Civil et Bâtiment, 3 rue Maurice Audin, F-69518 Vaulx-en-Velin, France

ABSTRACT

Total annoyance due to combined noises is still difficult to predict adequately. This scientific gap is an obstacle for noise action planning, especially in urban areas where inhabitants are usually exposed to high noise levels from multiple sources. In this context, this work aims to highlight potential to enhance the prediction of total annoyance. The work is based on a simulated environment experiment where participants performed activities in a living room while exposed to combined road traffic and industrial noises. The first objective of the experiment presented in this paper was to gain further understanding of the effects on annoyance of some acoustical factors, non-acoustical factors and potential interactions between the combined noise sources. The second one was to assess total annoyance models constructed from the data collected during the experiment and tested using data gathered *in situ*. The results obtained in this work highlighted the superiority of perceptual models. In particular, perceptual models with an interaction term seemed to be the best predictors for the two combined noise sources under study, even with high differences in sound pressure level. Thus, these results reinforced the need to focus on perceptual models and to improve the prediction of partial annoyances.

INTRODUCTION

Environmental noise remains a major concern. For instance, a recent French survey showed that more than 86% of French people are annoyed by noise at home [1]. Annoyance is one of the five health effects that the World Health Organization (WHO) considered in order to estimate the burden of disease due to environmental noise [2]. The densification of urban areas, activities generated by economic growth and motorized transport are some of the driving forces for environmental noise exposure and its health effects. Moreover, these driving forces lead to a multiplication of various combined noise exposure situations. For these situations, there is a lack of regulatory framework. Actually, the environmental noise directive [3], obliging the production of noise maps for European agglomerations, only considers noise sources separately, thus neglecting combined noise exposure. However when it comes to noise action planning, especially in urban areas, public authorities have to deal with multiple noise sources.

Annoyance due to combined noises has been investigated by researchers for more than 40 years [4]. However, knowledge remains incomplete in terms of: (1) assessment of potential interactions between combined noises and of (2) total annoyance prediction to better take into account these interactions [5,6,7]. Such knowledge is essential for planning effective and efficient noise control and abatement strategies for combined exposure situations [8].

From a methodological point of view, noise annoyance may be studied either *in situ* or under laboratory conditions. Laboratory studies are related to short-term annoyance assessed in an imaginary situation for short noise sequences, thus the corresponding judgments represent an annoyance potential [9]. *In situ* studies are related to long-term annoyance and the collected responses reflect actual annoyance as respondents are asked at their home to give a retrospective judgment [10]. Laboratory and *in situ* studies have their own advantages [11] and appear complementary: Laboratory conditions allow to mainly investigate the different acoustical factors (e.g., sound level, spectral content, *etc.*) and their potential interactions when combined noises are considered (e.g., [7]); *in situ* conditions allow the study of non-acoustical factors (e.g., noise sensitivity, fear of danger from the noise source, *etc.*, see [5,12,13] for reviews) encountered in real-life situations. Simulated environment experiments, either in laboratory (e.g., [14,15]) or in field (e.g., [16]), allow one to approach real-life situations when investigating the influence on annoyance of acoustical and non-acoustical factors under controlled conditions: Participants are placed in a room furnished like a typical living room, and they can perform activities while exposed to noise sequences. For example, the simulated environment is useful to study the influence of some non-acoustical factors on annoyance under controlled conditions such as the type of activity carried out during the noise exposure and its disturbance which may be more difficult to assess through surveys (e.g., [17]).

This paper intends to contribute to a better characterization of annoyance due to combined road traffic and industrial noises through a simulated environment experiment. For the combined noise exposures studied, the aims of the experiment presented in this paper were: (1) to investigate potential effects on annoyance of some acoustical factors, non-acoustical factors, and of potential interactions between the combined noise sources, (2) to assess total annoyance models from literature in terms of their goodness-of-fit (models were constructed from the data collected during the experiment) and in terms of their predictive power using *in situ* data gathered during a socio-acoustic survey [18,19,20].

First the experimental method is presented (Section 2). The investigation of potential effects on annoyance and of potential interactions between combined noises is then detailed (Section 3). Section 4 deals with the assessment of total annoyance models both in terms of goodness-of-fit and of predictive power. The results are then discussed (Section 5) and the main conclusions are drawn (Section 6).

EXPERIMENTAL METHOD

Noise Recording

The road traffic noise and the industrial noise under study corresponded to noise exposure in a residential neighborhood of a suburban city of Lyon. The inhabitants of this neighborhood lived close to a pharmaceutical plant and were respondents of an *in situ* survey carried out on total annoyance due to combined road traffic and industrial noises [18,19,20]. The survey area was located across from the industrial plant and delimited by two open roads. The industrial plant emitted a continuous and stationary industrial noise composed of cooling and air blowing device noises. At peak hours of road traffic, the industrial noise was mainly masked by the urban road traffic noise, whereas at off-peak hours without road traffic noise events, the continuous and steady industrial noise was unmasked.

All the noise recordings were made using the ORTF technique (Schoeps MSTC 64 microphones and BBG windfields), and stored on a portable recorder (TASCAM HD-P2, sampled at 44.1 kHz with an amplitude resolution of 24 bits). The ORTF couple was placed at a height of 1.5 m and at least at 2 m from any reflecting wall [21].

Recordings of vehicle pass-by noises were undertaken close to traffic lights along open roads including the ones along the survey area. The axis of the ORTF couple was placed perpendicularly to the road [22]. The recorded pass-by noises corresponded to various urban vehicle types (light vehicles, heavy vehicles, powered-two-wheelers and buses) in various driving conditions (acceleration, deceleration, constant speed). The recordings were carried out during the day. From these recordings, road vehicle pass-by noises with no masking noise were considered in order to be used for stimuli construction (*cf.* Section 2.2). The duration of the pass-by noises ranged from 3 s to 9 s.

For recording the industrial noise, the ORTF couple was placed in front of a house located in the surveyed area and facing the main noise sources of the industrial plant (cooling and air blowing devices [20]). The recording was carried out at night to be free of masking environmental noise.

Stimuli

Noise sequences of 7 min 30 s were constructed from the recorded noises. This duration was chosen considering similar simulated environment experiments previously published (e.g., [15,16]).

The industrial noise sequence was constructed using the recorded continuous and steady industrial noise. In order to allow the results of this current study to be tested with the data collected *in situ*[18,19,20], the industrial noise exposure level observed during the survey was considered. The A-weighted equivalent sound pressure level of the industrial noise, denoted by L_{ind}, was set in order to expose the experiment participants to indoor L_{ind} varying from 38 to 44 dB (A) in 2-dB (A) steps.

The road traffic noise sequence was constructed using the different recorded vehicle pass-by noises. This sequence had the following characteristics:

- Differences in A-weighted equivalent sound pressure level were considered between the vehicle pass-by noises that composed the road traffic noise sequence; they corresponded to mean differences observed *in situ* (*cf.* [23,24]) according to urban vehicle type (light vehicles, heavy vehicles, powered-two-wheelers and buses) and the driving conditions (acceleration, deceleration, constant speed);

- The considered composition of the urban road traffic corresponded to a traffic composition observed at peak hour (63% of light vehicles, 11% of heavy vehicles, 4% of buses, 22% of powered-two-wheelers);
- For the same reason as the one expressed for the industrial noise sequence, the A-weighted equivalent sound pressure level of the constructed road traffic noise sequence, denoted by L_{road}, was set in order to expose the experiment participants to indoor L_{road} varying from 44 to 53 dB (A) in 3-dB (A) steps.

The construction of combined road traffic and industrial noise sequences was based on the principle advised by Berglund and Nilsson [25] in order to study annoyance due to combined noises. The stimuli consisted of all possible combinations of the four noise levels of the road traffic noise sequence with the four noise levels of the industrial noise sequence (*i.e.*, a 4 × 4 matrix). Thus, the experiment involved 16 stimuli: 4 "Road traffic noise equivalent sound pressure level—L_{road}" × 4 "Industrial noise equivalent sound pressure level—L_{ind}".

Figure 1(a) shows the A-weighted sound pressure level $L_A(t)$ *versus* time for the combined noise sequence with industrial noise L_{ind} set at 44 dB (A) and road traffic noise L_{road} set at 47 dB (A). It displays peaks in L_A (t) due to the road vehicle pass-bys.

Figure 1(b) presents the frequency content of the same noise sequence. It displays the energy content of the steady industrial noise distributed over different stationary components which are illustrated by horizontal lines (*i.e.*, same energy content *versus* time):

- At low frequencies (energy content for frequencies up to 400 Hz),
- At middle frequencies (energy content for frequencies ranged from 500 Hz to 800 Hz),
- At higher frequencies (energy content for frequencies ranged from 900 Hz to 1500 Hz),
- Two narrower components around 1.7 kHz and 2.3 kHz respectively.

More details of this frequency content are displayed in Supplementary Figure S1. The energy content due to the different road vehicle pass-bys within the sequence was illustrated by vertical lines (*i.e.*, energy content for a large frequency range and only present during the vehicle pass-bys). This energy content is often distributed from 80 Hz up to frequencies higher than 6.4 kHz, depending on the vehicle type and on its driving condition. Such sound frequency content due to urban road vehicle pass by may be seen inSupplementary Figure S2.

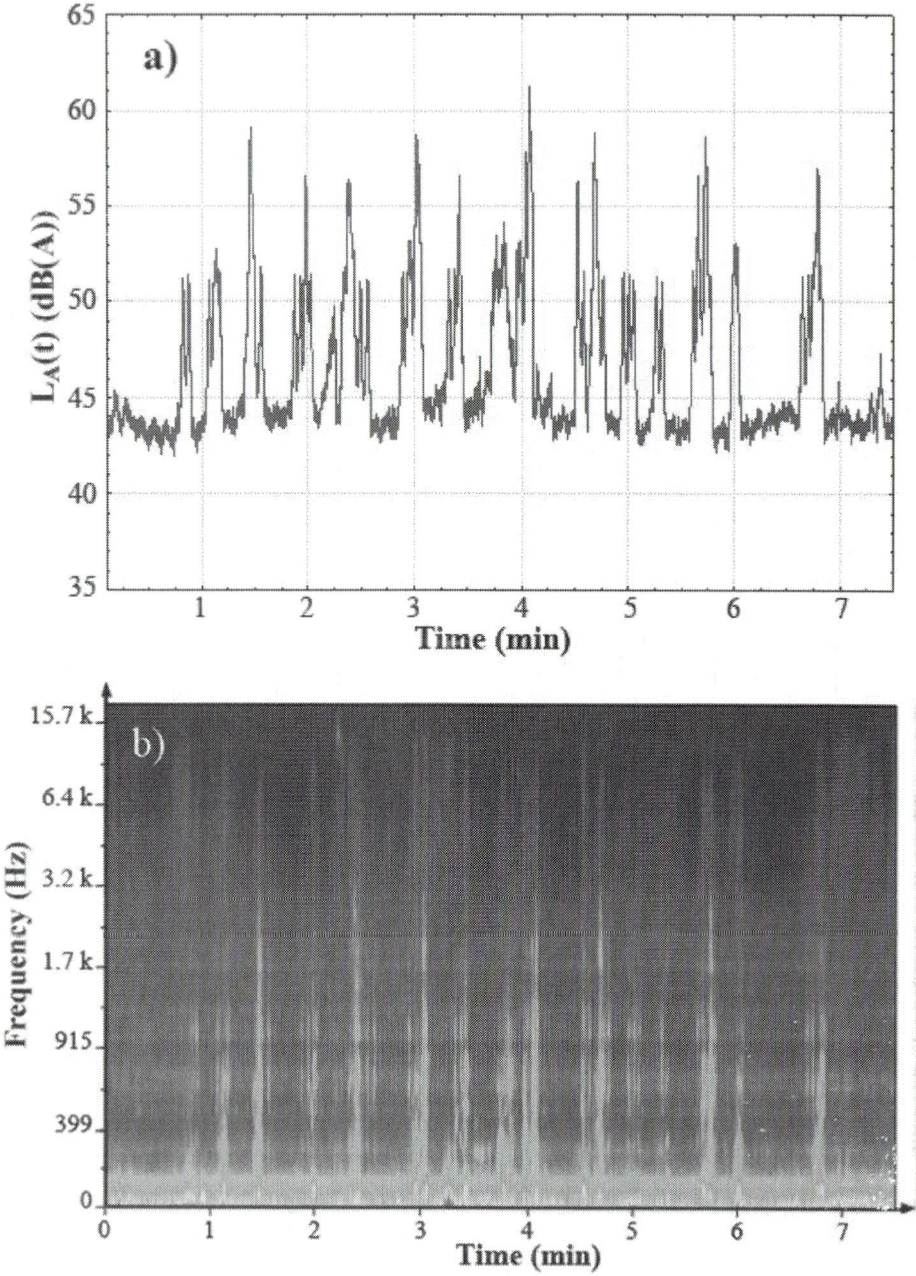

Figure 1. (**a**) A-weighted sound pressure level $L_A(t)$ *versus* time for the noise sequence combining the industrial noise set at 44 dB (A) with the road traffic

noise set at 47 dB (A); (**b**) The spectrogram for the noise sequence combining the industrial noise set at 44 dB (A) with the road traffic noise set at 47 dB (A).

During 57% of the combined noise sequence duration, the industrial noise was heard alone. During the other 43% of the sequence duration, the road traffic noise composed of different urban vehicle pass-bys masked the industrial noise. This was in agreement with the observation made *in situ* [18,19,20].

Experimental Setting

The experiment took place in a detached house located on the campus site of Vaulx-en-Velin, a suburban city of Lyon. The detached house was surrounded by a garden with trees and tall hedges. A residential road was in the vicinity at about 200 m. The residential urban sound environment around the detached house was rather quiet (percentile level L_{90} = 39 dB (A) during the day period). There would be human activities at certain periods of the day or wind in the tree leaves resulting in a higher outdoor background noise. Thus, attention was paid to carry out the experiment at quiet hours towards human activities and with appropriate weather conditions. During the experiment the window was kept partially opened, and the indoor A-weighted equivalent sound pressure level of the background noise in this condition was measured at 36 dB (A) (indoor percentile level L_{90}=32 dB (A)). For such condition, the industrial noise composed of cooling and air blowing device noises was always clearly distinguishable as it was the case in the survey area at periods of the day when no urban road vehicles were passing-by [18,19,20].

Two rooms with view on the garden and its hedges were used for the experiment. One room was furnished like a conventional living room and dedicated to hosting the participants during the experiment. The other room, adjacent to the first one, was dedicated to the control of the experiment.

The experimenter, placed in the control room, reproduced the stimuli in random order, using SoundForge©Sony software. The 2.1 reproduction system (2 Dynaudio Acoustics BM5A Active loudspeakers associated to a Dynaudio Acoustics BM9S subwoofer) was placed outside the living room in the garden as shown in Figure 2. The participants were seated on sofas around the coffee table. The central point between the participants was approximately the center of the coffee table. This central point faced the loudspeakers placed at a 1 m 20 height, forming an equilateral triangle of 5 m edge in agreement with Beck and Zacharov's recommendations [26].

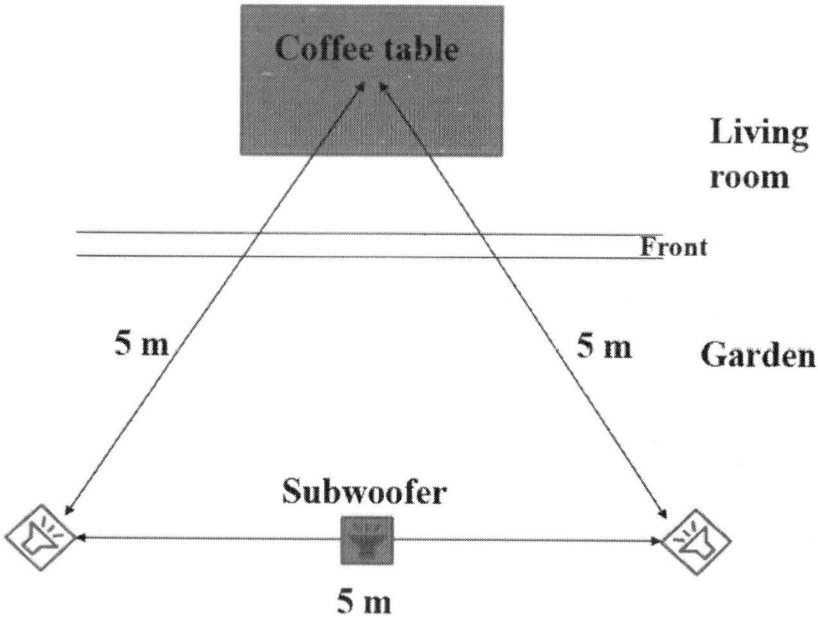

Figure 2. Experimental setting.

As the noise sequences were reproduced from outside to inside the living room with window partially-open, a frequency filtering was taken into account in the noise exposure of participants. The reproduced noise sequences were recorded inside at the central point to check the stimuli indoor levels defined inSection 2.2.

Procedure

Due to the experiment duration (3 h), the experiment was performed on two separate days. During each part (of about 1 h 30 each) participants were submitted to eight of the 16 noise sequences. Furthermore, four stimuli with combined road traffic and industrial noises were added in each part of the experiment in order to assess the homogeneous character of responses from one part of the experiment to the other, when the same stimuli were considered. First, participants had to read a paper with general instructions:

- They were told that they would be exposed to different sound environment sequences comprising industrial and road traffic sources, and that they were not allowed to close the window;
- They were proposed to imagine themselves at home with friends or colleagues, and that behind the garden's hedge, there was a crossroad and an industrial plant on the other side of the road;
- They were invited to perform a quiet and relaxing activity of their choice; examples of such activities were given: Reading, having a quiet conversation, playing cards; but they were not allowed to listen to music, give phone calls or watch videos.

Then, the experimenter went to the control room and played the first noise sequence. After each sequence, the participants were asked to answer a questionnaire. This questionnaire contained various questions:

- They were asked to write down the activity they performed during the exposure to the sound environment sequence;
- They had to assess how their activity was disturbed by the sound environment sequence on a continuous scale ranging from 0 "not at all disturbed" to 10 "extremely disturbed";
- They had to answer the question whether they wished to close the window because of the sound environment sequence through a dichotomous choice ("yes" or "no");
- They were asked to write which noise source they were able to identify during the sound environment sequence;
- They had to assess how much they were annoyed by the road traffic noise: "When you imagined yourself at home in presence of this sound environment sequence, did the road traffic noise annoy you?" (Question adapted from [10]). For this purpose, they were asked to use a continuous scale ranging from 0 "not at all annoyed" to 10 "extremely annoyed";
- They had to assess how much they were annoyed by the industrial noise: "When you imagined yourself at home in presence of this sound environment sequence, did the industrial noise annoy you?" (Question adapted from [10]). They were asked to give their ratings along a continuous scale ranging from 0 "not at all annoyed" to 10 "extremely annoyed";
- Finally, they had to assess how much they were annoyed by the overall sound environment sequence: "When you imagined yourself at home in presence of this sound environment sequence, did the overall noise, due to road traffic noise and industrial noise, annoy you?" (Question

adapted from [10]). They were asked to answer on a continuous scale ranging from 0 "not at all annoyed" to 10 "extremely annoyed".

At the end of the second part of the experiment, participants had to use three distinct continuous scales ranging from 0 "not at all to" 10 "extremely" in order to answer the following questions:

- "Would you say that you are sensitive to noise in general?"
- "Is for you a road infrastructure synonymous with danger?"
- "Is for you an industrial plant synonymous with danger?"

Participants

Fifty-one participants conducted the experiment (35 men, 16 women, mean age = 29, std. err. = 9.6 years). They were member staff and students from ENTPE recruited by email. They were told that we were looking for participants to a psychoacoustic test with relaxing activities in a dwelling room exposed to sound environment sequences with industrial and road traffic sources. All declared having normal hearing abilities. They were paid for their participation. Participants were separated in 12 groups of four or five participants each according to their availabilities.

Statistical Analysis

Statistical analyses, which included underlying hypotheses testing, were undertaken using STATISTICA software. First, correlation coefficients were calculated between annoyance ratings, noise sensitivity ratings and exposure noise levels in order to compare the results to findings from the *in situ*study [18,19] which combined noise situations were simulated in the current experiment. As the ratings were collected on a continuous scale during the experiment, Bravais-Pearson correlation coefficients r were calculated [27]. They are presented in Section 3.1.

Different analyses were carried out on different dependent variables collected during the experiment (e.g., activity disturbance, see [24] for further details). The analyses presented in Section 3 and Section 4related to annoyance were carried out to fulfill the two aims of the experiment presented in this paper.

The first aim dealt with the potential effects on annoyance of some acoustical factors, non-acoustical factors and interactions between the combined noise sources under study. The studied acoustical factors related to the noise exposure were indoor L_{ind} and L_{road}. The non-acoustical factors related to participants were the type of activity performed (denoted Ac), the noise sensitivity (denoted Se), the fear of danger from road infrastructure (denoted FR) and the fear of danger from industrial plant (denoted FI). The potential effects of these factors were studied on partial annoyance due to road traffic noise (denoted A_{road}), on partial annoyance due to industrial noise (denoted A_{ind}) and on total annoyance due to combined industrial and road traffic noises (denoted A_T).

In order to investigate the effects of L_{ind} and L_{road} (and their interaction) on A_{ind}, A_{road} and A_T, 2-way repeated measures analyses of variance (RM ANOVA) were undertaken [27]. In order to investigate the effects of Ac, Se, FI and FR on A_{ind}, A_{road} and A_T, analyses of variance (ANOVA) [27] were carried out. All these analyses are summarized in Table 1. The results related to the first aim of the experiment are presented in Section 3.

Table 1. Summary of ANOVA carried out to fulfill the first aim of the experiment.

Annoyance	**Independent Variables Under Consideration**	
	Acoustical Factor	**Non-Acoustical Factor**
Partial industrial noise annoyance (A_{ind})	L_{ind}	Fear from industrial plant (FI)
	L_{road}	Noise Sensitivity (Se)
		Activity performed (Ac)
	2-way RM ANOVA	1-way ANOVA
Partial road traffic noise annoyance (A_{road})	L_{ind}	Fear from road infrastructure (FR)
	L_{road}	Noise Sensitivity (Se)
		Activity performed (Ac)
	2-way RM ANOVA	1-way ANOVA
Total annoyance (A_T)	L_{ind}	Fear from road infrastructure (FR)
	L_{road}	Fear from industrial plant (FI)
		Noise Sensitivity (Se)
		Activity performed (Ac)
	2-way RM ANOVA	1-way ANOVA

The second aim of this work was to assess the goodness-of-fit of total annoyance models and their prediction power using data collected *in situ* [18,19,20]. To fulfil this aim, linear regression analyses were undertaken (*cf* Section 4).

EFFECTS OF INDEPENDENT VARIABLES ON ANNOYANCE

The homogeneous character of the ratings collected during each part of the experiment was checked using Student's t-tests on responses collected for the four common stimuli used in both parts of the experiment (*cf.* Section 2.4). As no significant differences were found in those responses collected in each part of the experiment, the data collected during both parts of the experiment were analyzed together.

Correlation Analyses

L_{ind} was significantly correlated with A_{ind} and A_T (respectively $r = 0.27, p < 0.001$; $r = 0.1, p < 0.005$). L_{road} was significantly correlated with A_{road} and A_T (respectively $r = 0.33, p < 0.001$; $r = 0.30, p < 0.001$). A_{road}, A_{ind} and A_T were significantly correlated with Se (r ranged from 0.24 to 0.38).

Effects of Acoustical Factors

On Partial Annoyance due to Industrial Noise—A_{ind}

The 2-way RM ANOVA showed that L_{ind} factor had a significant effect on partial annoyance A_{ind} due to industrial noise [$F(3,150) = 44.59$; $p < 0.001$]. According to the effect size measure η^2 (*i.e.*, the proportion of variance explained by this factor [27]), the effect was moderate ($\eta^2 = 9\%$). The L_{road} factor had no significant effect on A_{ind} [$F(3,150) = 1.32$; n.s. (*i.e.*, non-significant)]. Thus the increase in road traffic noise level had no influence on the annoyance due to industrial noise when heard within the combined sequence. In other words, the partial annoyance due to industrial noise was not influenced by the presence of road traffic noise.

On Partial Annoyance due to Road Traffic Noise—A_{road}

The 2-way RM ANOVA highlighted that L_{road} factor had a significant effect on partial annoyance A_{road} due to road traffic noise [$F(3,150) = 61.48$; $p < 0.001$] with a moderate effect size measure ($\eta^2 - 11\%$). The L_{ind} factor had no significant effect on A_{road} [$F(3,150) = 0.41$; n.s.]. Thus a similar conclusion as the one concerning A_{ind} could be drawn: The partial annoyance due to road traffic noise was not influenced by the presence of industrial noise.

On Total Annoyance due to Combined Industrial and Road Traffic Noises—A_T

The 2-way RM ANOVA showed that both factors L_{road} and L_{ind} had a significant effect on total annoyance A_T with moderate and weak effect size measure respectively ([$F(3,150) = 51.88$; $p < 0.001$; $\eta^2 = 9\%$] and

[F(3,150) = 9.17; $p < 0.001$; $\eta^2 = 2\%$] respectively). These results showed that the road traffic noise had a stronger effect on total noise annoyance in comparison with the industrial noise. This was due to the high difference in noise levels between these combined noises. But the results also showed that for the noise level range under consideration, the industrial noise had a significant weak influence on total annoyance. This was explained by the intermittent character of the road traffic noise and the steady character of industrial noise which was present between urban road vehicle pass-bys (*cf.* Figure 1).

The 2-way RM ANOVA also highlighted a significant effect on total annoyance of the interaction between the two factors with weak effect size measure [F(9,450) = 3.14; $p < 0.001$; $\eta^2 = 2\%$]. Figure 3illustrates this interaction.

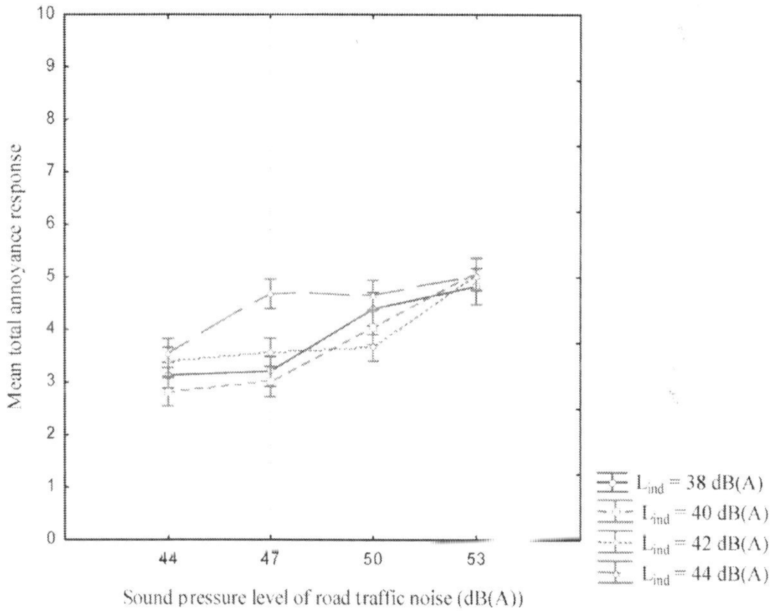

Figure 3. Mean total annoyance responses in function of factor L_{road} and categorized by each level of factor L_{ind}. Error bars represent the standard error.

The effect on total annoyance of the interaction between the combined noise source levels was due to the fact that at the higher level of industrial noise equal to 44 dB (A), the total annoyance did not increase anymore when the road traffic noise level increased from 47 dB (A) up to 53 dB (A).

Effects of Non-acoustical Factors

The activity performed by participants during each noise sequence and gathered after each noise sequence belonged to one of these four activity types: conversation (42% of the performed activities), individual concentration such as reading or crossword puzzle (32% of the performed activities), party game such as card game or jackstraws (25% of the performed activities) or just having a rest (only 1% of the performed activities).

The 1-way ANOVA (*cf.* Table 1) highlighted that each independent variable had a significant effect on the dependent variable under consideration. Table 2 details the significant effect of each main factor for each 1-way ANOVA performed. One can notice that values of the effect size measure η^2 were different. Among all the studied independent variables related to non-acoustical factors, the activity performed (Ac) explained the least amount of variance observed in the different dependent variables (η^2 was ranged from 2% to 9% according to the dependent variable under consideration). The fear of danger from industrial plant (FI) explained between 8% and 13% of the variance. Then the fear of danger from road infrastructure (FR) explained between 10% and 15% of the variance. Finally, noise sensitivity (Se) explained the greatest amount of variance observed in the different dependent variables (η^2 was ranged from 10% to 21%).

Table 2. 1-way ANOVA results.

Annoyance	Independent Variable	Degree of Freedom	F	*p*-Value	η^2 Effect Size Measure
A_{ind}	FI	9	6.52	<0.001	0.08
	Se	10	8.51	<0.001	0.10
	Ac	3	5.87	<0.001	0.02
A_{road}	FR	11	8.43	<0.001	0.10
	Se	10	17.87	<0.001	0.18
	Ac	3	25.80	<0.001	0.09
A_T	FI	9	13.21	<0.001	0.13
	FR	11	12.67	<0.001	0.15
	Se	10	21.49	<0.001	0.21
	Ac	3	20.12	<0.001	0.07

A_{ind}: Partial annoyance due to industrial noise; A_{road}: Partial annoyance due to road traffic noise; A_T: Total annoyance, FI: Fear of danger from industrial plant; FR: Fear of danger from road infrastructure; Se: Noise sensitivity; Ac: Activity performed.

As already mentioned, the 1-way ANOVAs showed that the activity performed (Ac) explained the least amount of variance (between 2% and 9% according to the dependent variable under consideration) but it was important to analyze the different quiet activities proposed to the participants in terms of their different influence on the reported noise annoyance. For each annoyance dependent variable, Tukey's HSD test [27] revealed that participants having a rest were more annoyed during the experiment than participants playing a party game. The Tukey's HSD test also revealed that except for partial annoyance due to industrial noise, participants playing a party game were less annoyed than participants having a conversation. The participants having a conversation were less annoyed than participants performing an individual activity. Figure 4 displays the mean responses of the dependent variables *versus* the activity performed by the participants. Due the weak proportion of participants having a rest, the standard error for this activity was the greatest.

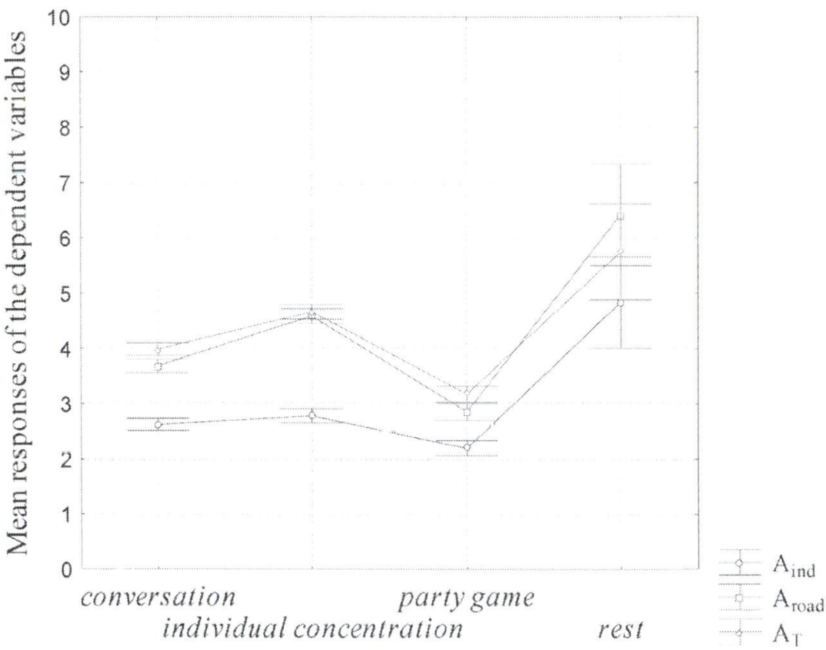

Figure 4. Mean responses of the dependent variables *vs.* the type of activity (Ac). Error bars represent the standard error. A_{ind}: Partial annoyance due to industrial noise. A_{road}: Partial annoyance due to road traffic noise. A_T: Total annoyance.

Comparison of Partial Annoyance and Total Annoyance

In order to investigate potential interactions between the two combined noise sources, related to perceptual phenomena (e.g., paradox effect [6]), and comparison of partial and total annoyances was performed using paired *t*-tests.

These comparisons showed that total annoyance was generally close to the highest partial annoyance. Actually for only one combination among the 16 under study, a synergetic effect was observed, *i.e.*, the total annoyance was significantly higher than the highest partial annoyance. For the other 15 combinations studied, there was no significant difference between the total annoyance and the highest partial annoyance. Thus, a strongest component phenomenon was mainly highlighted. It could then be expected that the strongest component model proposed in literature [28] would be appropriate to predict the total annoyance due to the combined industrial and road traffic noises. Comparison of the two partial annoyances showed that the strongest component was mainly the partial annoyance due to the road traffic noise (A_{road} was superior to A_{ind} in 14 combinations on 16).

TOTAL ANNOYANCE MODEL ASSESSMENT

The Models

The second aim of the experiment concerned the assessment of total annoyance models from literature: (1) in terms of their goodness-of-fit when linear regression models were built from the data collected during the simulated environment experiment, and (2) in terms of their predictive power using *in situ*collected data during a survey [18,19,20].

The total annoyance models were selected according to the independent variables available from the survey [18,19,20] in order to assess their predictive power. These *in situ* data were partial and total annoyance ratings and the values of the day-evening-night level L_{den} for each single noise exposure and for the global noise due to the combined noise sources [18,19,20]. The annoyance data were collected from 99 respondents [18] (50 men and 49 women; mean age = 45.9 years with SD = 17.9; length of residence = 11.6 years with SD = 10.3; Working 45.5%; 8% worked or had a relative who worked on the industrial site). The annoyance ratings available [20] for testing annoyance models were gathered using the questions recommended by the ISO standards [10] and a continuous scale with two numerical labels 0 and 10 at both ends [18]. Eight total annoyance models from the literature were assessed (see for example [7,18,19] for an extensive description and discussion of these models):

- Four psychophysical models using independent variables constructed from L_{Aeq} (*cf.* Table 3):
- Energy summation model [29], based on the overall sound pressure level (L_T) calculated as an energy summation of L_{Aeq} of each combined noise;
- Energy difference model [29], based on the overall sound pressure level (L_T) and the absolute difference between L_{Aeq} of each combined noise to account for possible interaction;
- Independent effect model [29], based on L_{Aeq} of each combined noise;
- Weighted summation model [30], based on the total rating sound level (Lt) defined by:

$$L_t = k.log\left(10^{\frac{L_{ref}}{k}} + 10^{\frac{L_1+P_1}{k}}\right) \tag{1}$$

where k is a parameter that optimizes the model, L_{ref} is the L_{Aeq} of the source chosen as the reference source in the combination, L_1 is the L_{Aeq} of the other source of the combination, and $P_1$the parameter calculated to obtain the equally annoying source reference level and given by:

$$P_1 = \frac{b_1 - b_{ref} + (a_1 - a_{ref})\,L_1}{a_{ref}} \tag{2}$$

where a_{ref}, b_{ref}, a_1 and b_1 were coefficients of linear regressions between L_{Aeq} of each combined noise source and their respective annoyance:

$$A_{ref} = a_{ref}\,L_{ref} + b_{ref} \tag{3}$$

$$A_1 = a_1\,L_1 + b_1 \tag{4}$$

- Four perceptual models using independent variables constructed from mean partial annoyance ratings (*cf.* Table 3):
- Strongest component model [28];

- Vector summation model [28] using an angle α that may allow to account for interaction and to optimize the model;
- Linear regression model [31];
- Mixed model [18,19] based on the absolute difference between the partial annoyances to account for possible interaction.

Goodness-of-Fit of the Total Annoyance Models Built from the Experiment Data

Table 3 summarizes the results of regression analyses performed for the assessment of the total annoyance models in terms of goodness-of-fit. The models were built linear regression analyses carried out between total annoyance responses (A_T) collected during the simulated environment experiment, and the experiment independent variables (L_{Aeq} for psychophysical models and mean partial annoyances for perceptual models, respectively). The higher the determination coefficient (R^2) and the lower the standard error of the estimate (std. err.), the better the goodness-of-fit.

Table 3. Total annoyance model assessment in terms of goodness-of-fit.

Model		Regression Equation	R^2	Std.err.
Psychophysical models	Energy summation	$A_T = 0.25L_T{}^{d} - 8.12$	0.82	0.35
	Energy difference	$A_T = 0.30\mathit{(1.08)}L_T{}^{d} - 0.05\mathit{(-0.22)}\lvert L_{road}-L_{ind}\rvert^{a} - 9.82$	0.82	0.34
	Independent effects	$A_T = 0.22\mathit{(0.85)}L_{road}{}^{d} + 0.08\mathit{(0.21)}L_{ind}{}^{a} - 10.36$	0.80	0.36
	Weighted summation (k = 7)	$A_T = 0.31L_t{}^{d} - 10.24$	0.84	0.33
Perceptual models	Strongest component	$A_T = 0.93\max(A_{road},A_{ind})^{d} + 0.44$	0.88	0.29
	Vector summation (α = 106°)	$A_T = 1.05\sqrt{(A_{road}{}^2 + A_{ind}{}^2 + 2A_{road}A_{ind}\cos\alpha)}^{d} - 0.18$	0.99	0.10
	Linear regression	$A_T = 0.81\mathit{(0.85)}A_{road}{}^{d} + 0.48\mathit{(0.41)}A_{ind}{}^{d} - 0.29$	0.98	0.13
	Mixed	$A_T = 0.64\mathit{(0.67)}A_{road}{}^{d} + 0.62\mathit{(0.53)}A_{ind}{}^{d} + 0.19\mathit{(0.21)}\lvert A_{road}-A_{ind}\rvert^{a} - 0.27$	0.98	0.12

R^2: The determination coefficient. Std. err.: The standard error of the estimate. Numbers in italics between brackets correspond to the standardized regression coefficient. [a] n.s. [b] $p < 0.05$. [c] $p < 0.01$. [d] $p < 0.001$. L_{road}: The A-weighted equivalent sound pressure level of the road traffic noise. L_{ind}: The A-weighted equivalent sound pressure level of the industrial noise. L_T: The A-weighted overall sound pressure level. L_t: Total rating sound level calculated with the parameter k [30]. A_{ind}: Partial annoyance due to industrial noise; A_{road}: Partial annoyance due to road traffic noise. A_T: Total annoyance.

Two models involve the calculation of specific parameters. For the weighted summation model [30], the industrial noise was considered as the reference source, and the best goodness-of-fit was obtained with the parameter k equal to 7 for the calculation of the total rating sound level L_t. For the vector summation model [28], an angle $\alpha = 106°$ allowed the model to be optimized.

Concerning the psychophysical models, Table 3 showed that the weighted summation model led to the best calculation of total annoyance, although the differences with the other psychophysical models were small. Figure 5 shows that the weighted summation model was a good model with no tendency to overestimate or underestimate total annoyance, even if the spread of data points appeared generally large.

Concerning the perceptual models, Table 3 shows that the strongest component model yielded the worst calculation of total annoyance, whereas the other three models better succeeded. The result relative to the strongest component model was not expected as the strongest component phenomenon was mainly highlighted for the different combined noise situations studied (*cf.* Section 3.4).

Furthermore, Table 3 shows that perceptual models provided a better calculation of total annoyance than psychophysical models. This result is illustrated by Figure 5 displaying the best psychophysical model (the weighted summation model), and the best perceptual model (the vector summation model). As a matter of fact, the spread of data points in the scatter plot concerning the vector summation model was smaller than the one concerning the weighted summation model.

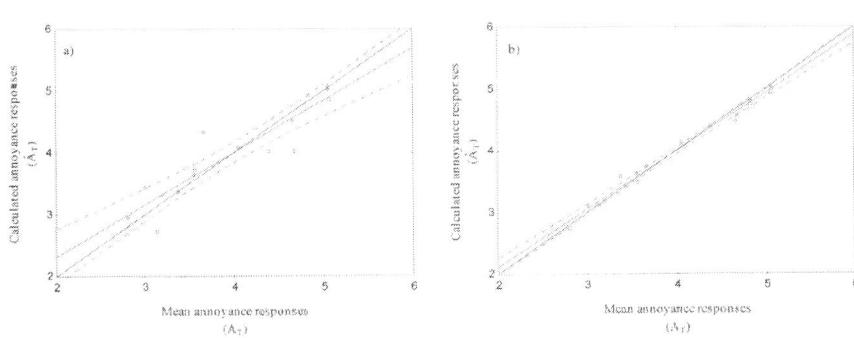

(a) (b)

Figure 5. Calculated total annoyance responses ($\hat{A}_T$) *versus* mean annoyance responses collected during the simulated environment experiment (A_T); (**a**) the weighted summation model ($k = 7$); (**b**) the vector summation model ($\alpha = 106°$).

Total Annoyance Model Testing Using Data Collected in Situ

The three best models highlighted in the previous section (*i.e.*, the three perceptual models with the best goodness-of-fit: The vector summation model, the linear regression model and the mixed model) were considered to be assessed in terms of their predictive power. The partial annoyance responses A_{road} and A_{ind}, collected during the survey [18,19,20], were used in the equations built from the data collected during the simulated environment experiment (*cf.* Table 3). Thus predicted annoyance responses were obtained ($\hat{A}_T$').

The predictive power of these equations was assessed using r, a and b: r is the Bravais-Pearson coefficient correlation between A_T (total annoyance collected *in situ*) and $\hat{A}_T$' (predicted total annoyance), a and b are the slope and intercept of the regression line between A_T and $\hat{A}_T$'.

Table 4 and Figure 6 show that the vector summation and mixed models performed best, and led to quite similar results. In Figure 6, data points were more closely clustered to the regression line using the vector summation and the mixed model relative to the linear regression model.

Table 4. Total annoyance model testing results.

Model	r [d]	a	b
Vector summation	0.96	1.02	0.47
Linear regression	0.86	0.92	0.26
Mixed model	0.92	0.99	0.23

[d] $p < 0.001$.

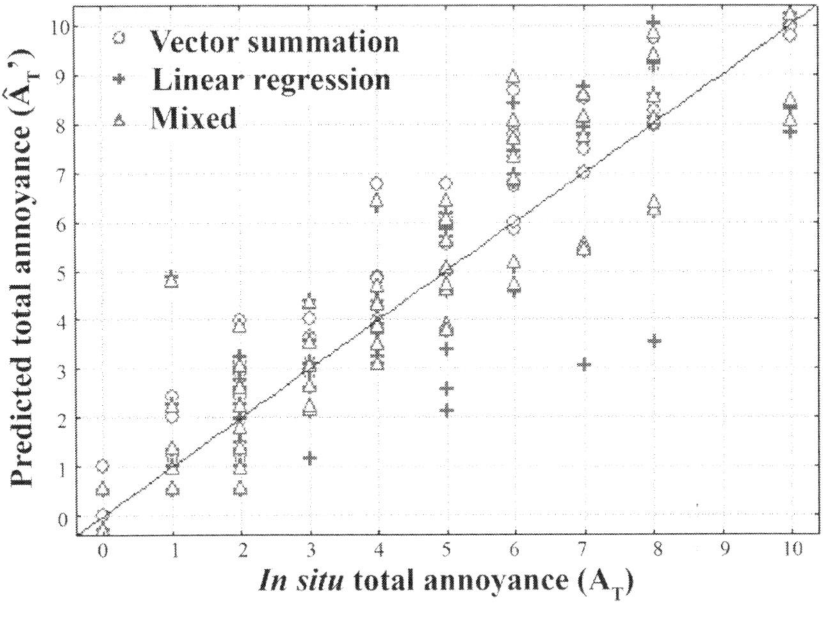

Figure 6. Scatter plot representing the results of total annoyance model testing.

DISCUSSION

The correlation analyses carried out on the data collected during the experiment highlighted significant correlation coefficients between:

- Road traffic noise level and partial road traffic noise annoyance ($r = 0.33$, $p < 0.001$),
- Industrial noise level and partial industrial noise annoyance ($r = 0.27$, $p < 0.001$),
- Annoyances and noise sensitivity collected on 1-item scale (r ranged from 0.24 to 0.38).

The results obtained for annoyance due to road traffic noise were in agreement with those of different field studies (e.g., [13,32]), in terms of either correlation coefficient between annoyance and noise level, or correlation coefficient between annoyance and noise sensitivity. Furthermore, the simulated environment experiment dealing with annoyance due to combined industrial and road traffic noises allowed to provide results in agreement with the findings of the socio-acoustic survey [18,19] from which the current experiment was based on. Actually,

Pierrette *et al.* [18,19] found a significant correlation coefficient, equal to 0.38, between road traffic noise level and partial road traffic noise annoyance. They found a significant correlation coefficient, equal to 0.27, between industrial noise level and partial industrial noise annoyance. Pierrette *et al.* also found significant correlation coefficient between annoyances and noise sensitivity, the correlation coefficients were ranged from 0.2 to 0.27 [18,19].

Concerning the influence of the activity performed during the experiment on annoyance responses, we found similar results as the one highlighted by Öhrström *et al.* [32] in their field study dealing with road traffic noise exposure: Activities involving speech communication were less impacted by noise compared to concentration and relaxation activities. Thus, the experiment simulated real-life situations and combined noise exposures allowed to provide results in agreement with findings of field studies.

The ANOVA carried out in the current work showed that the exposure to road traffic noise had no significant effect on partial annoyance due to the industrial noise, and that the exposure to industrial noise had no significant effect on partial annoyance due to road traffic noise. Thus the partial annoyances were not influenced by the presence of the other noise in the combination, contrary to inhibition observed between combined transportation noises (e.g., [4]). One possible explanation lies in the fact that for the considered noise sequences, the differences in sound pressure level between the highly varying road traffic noise and the steady and continuous industrial noise were important. Due to these differences, road vehicle noises, when present, masked almost completely the industrial noise. This masking effect had also been observed *in situ* for these two combined noises [18,19] and led to a description of the steady industrial noise as an intermittent noise by the residents as they only heard it when no road traffic pass-by noise events were present.

The 2-way RM ANOVA showed the strongest effect of L_{road} on the total annoyance as it was expected, but this ANOVA also showed that despite the relatively important differences in L_{Aeq} between the two combined noises, significant weak effects of L_{ind} and of the interaction between L_{road} and L_{ind} existed on total annoyance A_T.

The comparison of total annoyance and partial annoyances highlighted mainly a strongest component phenomenon, and occasionally but significantly, a synergetic effect. These results were in agreement with the significant standardized regression coefficients assigned to variables based on A_{road} and A_{ind} in the perceptual models (*cf.* Table 3). These results were also in agreement with observations carried out *in situ*by Pierrette *et* *al.* [18,19]: Despite a lower level of the industrial noise in the combination, the corresponding partial annoyance was non-negligible as only 51% of people surveyed found the industrial noise less annoying than the road traffic noise, and 27% of them found both noises equally annoying.

The comparison of total annoyance models (*cf.* Table 3 and Figure 5) showed that perceptual models are superior to psychophysical models concerning total annoyance calculation, which was congruent with literature findings (e.g., [18,19]). Moreover, the results presented in Table 3 pointed out that the strongest component model was not the best model whereas a strongest component phenomenon was mainly highlighted for the studied combined noise situations. Both *in situ* and laboratory noise annoyance studies had also pointed out such limitations of this model [7,18,19,31]. The main drawbacks of the strongest component model lie in the fact that it does not allow to take into account (1) possible influence of variations in the partial annoyance of the least annoying noise on the total annoyance [31], and (2) possible interactions between combined noises [7,18,19]. This result was confirmed in the experiment using important differences in noise levels between the two combined noises which mainly led to the strongest component phenomenon.

The total annoyance model testing consisted in injecting the data of an *in situ* survey into the model equations built from the simulated environment experiment. The comparison of perceptual models highlighted that the vector summation and the mixed models allowed a better prediction of *in situ* total annoyance than the linear regression model (*cf.* Table 4 and Figure 6). This result was congruent with results obtained by Pierrette *et al.* [18,19]. This showed that it was necessary to propose models allowing possible interactions between noises to be taken into account (the angle α used in the vector summation model and the absolute difference term used in the mixed model were examples of such possibility).

This work pointed out that the simulated environment experiment was useful to contribute to enhance total noise annoyance model. Actually, this experiment which simulated combined industrial and road traffic noises observed *in situ* allowed the construction of perceptual total annoyance models with a good predictive power.

This work also highlighted the limit due to *in situ* data available for model testing. Actually no annoyance models including both acoustical and non-acoustical variables (e.g., multilevel regression model using acoustical variables and noise sensitivity [33]) were built and tested in this work.

This suggests encouraging surveys which allow to collect *in situ* data available for model testing. Such surveys might be carried out with common procedures in order to collect annoyance and influential non-acoustical factors on continuous 1-item scale as recently done by Ecotière *et al*. [34] for a survey dealing with transportation noise.

Such suggestions would be useful to improve the actual dose-effect relationships, used to model the impact on health of noise exposure in terms of noise annoyance, which is a challenge identified by WHO [2]. Furthermore, adequate dose-effect relationships used in a perceptual model of total annoyance could thus help noise action planners, especially in urban areas where inhabitants are exposed to multiple noise sources. As a matter of fact, the European Environment Agency [35] in its guideline recommends to use a psychophysical model based on actual dose-effects relationships which drawbacks are clearly identified [2].

CONCLUSIONS

This work dealt with total annoyance due to industrial and road traffic noises using a simulated environment experiment. During the experiment, participants were in conditions approaching real-life situation. They were exposed to combined industrial and road traffic noise sequences that simulated the noise environment of a suburban city of Lyon where a socio-acoustic survey was carried out.

The main results were the following:

- Concerning acoustical factors, no inhibition effect on partial annoyance due to the presence of the other noise were highlighted. Also, due to the difference in sound pressure levels between the two combined noises, the road traffic noise level had a greater influence on total annoyance than the industrial noise level;
- All of the considered non-acoustical factors had significant effects on the studied dependent variables;
- The comparison of partial and total annoyances mainly highlighted a strongest component effect;
- The assessment and comparison of total annoyance models confirmed the superiority of perceptual models over psychophysical models;

- The testing of the total annoyance models using *in situ* data highlighted the superiority of the perceptual models with interaction term. Such result is interesting as such models allow interaction between combined noises to be taken into account even in the studied case of total annoyance mainly governed by strongest component phenomenon.

Thus, these results reinforced the need to focus on perceptual models and to improve the prediction of partial annoyances. On the long term, this could also help to improve noise action planning especially in urban areas where multiple noise source environments are most often experienced by inhabitants.

SUPPLEMENT

Supplementary File 1:

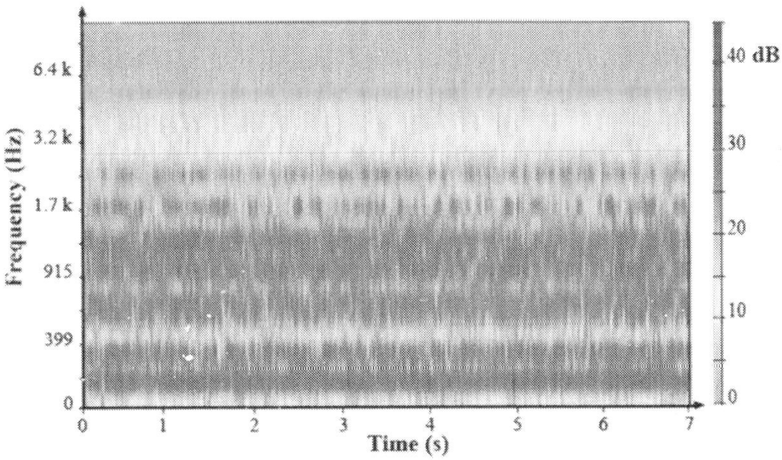

Figure S1. Spectrogram for an excerpt (7s) of the studied stationary industrial noise: zoom on its frequency content.

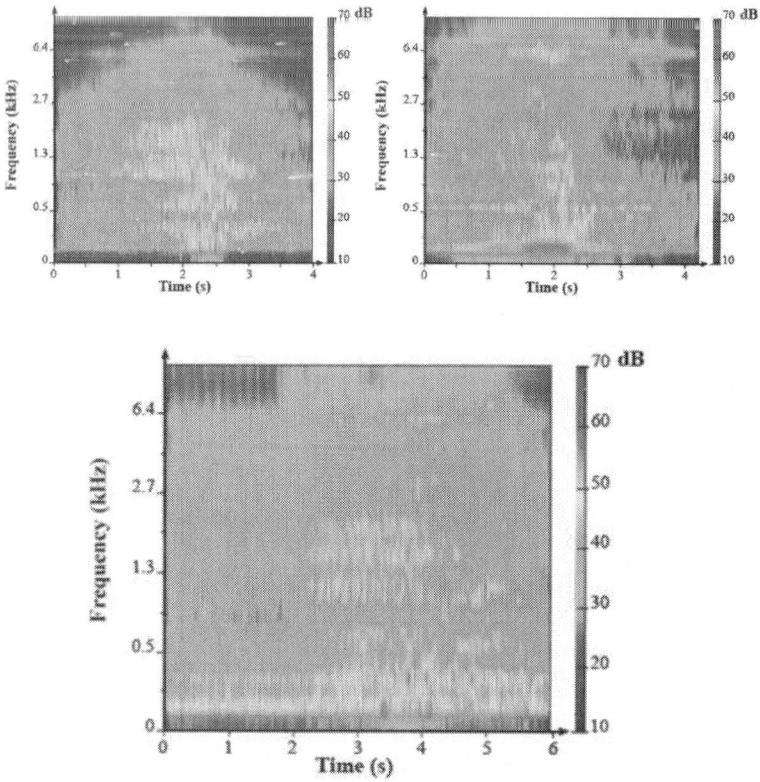

Figure S2. Spectrogram for excerpts of the studied urban road traffic noise: zoom on the spectral content due to urban road vehicle pass-by (a light vehicle at constant speed: top left, a powered-two-wheeler in deceleration: top right, a heavy vehicle in acceleration: bottom left).

ACKNOWLEDGMENTS

This work was performed within the framework of the Labex CeLyA of Université de Lyon, operated by the French National Research Agency (ANR-10-LABX-0060/ANR-11-IDEX-0007).

REFERENCES

1. Bendavid, R.; Chasles-Parot, M. *Les Français et les Nuisances Sonores (French and Noise Nuisances)*; Institut français d'opinion publique (IFOP): Paris, France, 2014; p. 24, (in French).
2. Fritschi, L.; Brown, L.; Kim, R.; Schwela, D.; Kephalopoulos, S. *Burden of Disease from Environmental Noise: Quantification of Healthy Life Years Lost in Europe*; World Health Organization, Regional Office for Europe: Copenhagen, Denmark, 2011; p. 128.
3. European Parliament and Council. *Guideline 2002/49/CE Relative to the Assessment and Management of Environmental Noise*; European Parliament, Council of the European Union: Brussels, Belgium, 2002.
4. Bottom, C.G. A social survey into annoyance caused by the interaction of aircraft noise and traffic noise. *J. Sound Vib.* **1971**, *19*, 473–476.
5. Marquis-Favre, C.; Prémat, E.; Aubrée, D. Noise and its effects—A review on qualitative aspects of sound. Part II: Noise and annoyance. *Acta Acust. United Acust.* **2005**, *91*, 626–642.
6. Alayrac, M.; Marquis-Favre, C.; Viollon, S. Total annoyance from an industrial noise source with a main spectral component combined with a background noise. *J. Acoust. Soc. Amer.* **2011**, *130*, 189–199.
7. Morel, J.; Marquis-Favre, C.; Viollon, S.; Alayrac, M. A laboratory study on total annoyance due to combined industrial noises. *Acta Acust. United Acust.* **2012**, *98*, 286–300.
8. Lercher, P. Combined noise exposure at home. In *Encyclopedia of Environmental Health*; Elsevier: Burlington, MA, USA, 2011; pp. 764–777.
9. Flindell, I.H. Community Response to Multiple Noise Sources. Ph.D. Thesis, Southampton University, Southampton, UK, 1982.
10. ISO 15666. In *Acoustics—Assessment of Noise Annoyance by Means of Social and Socio-Acoustic Surveys*; International Organization for Standardization (ISO): Geneva, Switzerland, 2003.
11. Rice, C.G. Human response effects of impulse noise. *J. Sound Vib.* **1996**, *190*, 525–543.
12. Klaeboe, R. Noise and health: Annoyance and interference. *Encycl. Environ. Health* **2011**, *2011*, 152–163.
13. Job, R.F.S. Community response to noise: A review of factors influencing the relationships between noise exposure and noise reaction. *J. Acoust. Soc. Amer.* **1988**, *83*, 991–1001.
14. Rice, C.G. Development of cumulative noise measure for the prediction of general annoyance in an average population. *J. Sound Vib.* **1977**, *52*, 345–364.

15. Ota, O.; Yokoshima, S.; Tamura, S. The metrics of mixed traffic noise: Results of simulated environment experiments. In Proceedings of the 9th International Congress on Noise as a Public Health Problem (ICBEN), Foxwoods, CT, USA, 2008; p. 8.
16. De Coensel, B.; Botteldooren, D.; de Muer, T.; Lercher, P.; Berglund, B.; Nilsson, M.E. Experimental investigation of noise annoyance caused by high-speed trains. *Acta Acust. United Acust.* **2007**, *93*, 589–601.
17. Kurra, S.; Morimoto, M.; Maekawa, Z.I. Transportation noise annoyance—A simulated-environment study for road, railway and aircraft noises; part 2: Activity disturbance and combined results. *J. Sound Vib.* **1999**, *220*, 279–295.
18. Pierrette, M.; Marquis-Favre, C.; Morel, J.; Rioux, L.; Vallet, M.; Viollon, S.; Moch, A. Noise annoyance due to industrial and road traffic combined noises: A survey and a total annoyance model comparison. *J. Environ. Psych.* **2012**, *32*, 178–186.
19. Pierrette, M.; Marquis-Favre, C.; Morel, J.; Rioux, L.; Vallet, M.; Viollon, S.; Moch, A. Corrigendum to Noise annoyance due to industrial and road traffic combined noises: A survey and a total annoyance model comparison. *J. Environ. Psych.* **2012**, *32*, 285.
20. EDF; ENTPE; ORELIA. *Rapport Final du Projet NOISI-EXPO (Final Report of NOISI-EXPO Project)*; Agence de l'Environnement et de la Maîtrise de l'Energie: Sophia-Antipolis, France, 2010; p. 243, (in French).
21. AFNOR NF S31–085. In *Acoustique—Caractérisation et Mesurage du Bruit dû au Trafic Routier (Acoustics: Characterization and Measurement of Road Traffic Noise)*; Groupe AFNOR: Saint-Denis, France, 2002; (in French).
22. Morel, J.; Marquis-Favre, C.; Dubois, D.; Pierrette, M. Road traffic in urban areas: A perceptual and cognitive typology of pass-by noises. *Acta Acust. United Acust.* **2012**, *98*, 166–178.
23. Klein, A.; Marquis-Favre, C.; Weber, R.; Trollé, A. Spectral and modulation indices for annoyance-relevant features of urban road single-vehicle pass-by noise. *J. Acoust. Soc. Amer.***2015**, *137*, 1238 1250.
24. Morel, J. Caractérisation Physique et Perceptive du Bruit Routier Urbain pour la Définition D'indicateurs de Gêne Sonore en Situation de Mono-exposition et de Multi-exposition en Présence de Bruit Industriel (Physical and Perceptual Characterization of Road Traffic Noise for the Determination of Annoyance Indicators in Single and Combined Noise Exposure in the Presence of Industrial Noise). Ph.D. Thesis, Ecole Nationale des Travaux Publics de l'Etat, Vaulx-en-Velin, France, 2012. (in French).
25. Berglund, B.; Nilsson, M.E. Empirical issues concerning annoyance models for combined community noise. In *Inter Noise 97, Help Quiet the World for a Higher Quality of Life*; Augustinovicz, F., Ed.; Noise Control Foundation: Poughkeepsie, NY, USA, 1997; Volume 2, pp. 1053–1058.

26. Beck, S.; Zacharov, N. *Perceptual Audio Evaluation—Theory, Method and Application*; Wiley: New York, NY, USA, 2006; p. 442.
27. Howell, D.C. *Statistical Methods for Psychology*, 4th ed.; Duxbury: Belmont, CA, USA, 1997; p. 821.
28. Berglund, B.; Berglund, U.; Goldstein, M.; Lindvall, T. Loudness (or annoyance) summation of combined community noises. *J. Acoust. Soc. Amer.* **1981**, *70*, 1628–1634.
29. Taylor, S.M. A comparison of models to predict annoyance reactions to noise from mixed sources.*J. Sound Vib.* **1982**, *81*, 123–138.
30. Vos, J. Annoyance caused by simultaneous impulse, road-traffic, and aircraft sounds: A quantitative model. *J. Acoust. Soc. Amer.* **1992**, *91*, 3330–3345.
31. Botteldooren, D.; Verkeyn, A. Fuzzy models for accumulation of reported community noise annoyance from combined sources. *J. Acoust. Soc. Amer.* **2002**, *112*, 1496–1508.
32. Öhrström, E.; Skånberg, A.; Svensson, H.; Gidlöf-Gunnarsson, A. Effects of road traffic noise and the benefit of access to quietness. *J. Sound Vib.* **2006**, *295*, 40–59.
33. Trollé, A.; Marquis-Favre, C.; Klein, A. Short-term annoyance due to tramway noise: Determination of an acoustical indicator of annoyance via multilevel regression analysis. *Acta Acust. United Acust.* **2014**, *100*, 34–45.
34. Ecotière, D.; Champelovier, P.; Marquis-Favre, C.; Morel, J.; Olny, X.; Philipps-Bertin, C.; Vincent, B. *Multi-exposition au Bruit des Transports: Application et Evaluation d'une Méthode Opérationnelle D'identification de la Gêne en Situation de Multi-exposition (Combined Transportation Noise Sources: Assessment of a Practical Method to Predict Total Annoyance)*; Ministère de l'Ecologie, du Développement Durable et de l'Energie: Paris, France, 2014; p. 106, (in French).
35. European Environment Agency. *Good Practice Guide on Noise Exposure and Potential Health Effects*; European Environment Agency: København K, Denmark, 2010, p. 40.

CITATION

Catherine Marquis-Favre and Julien Morel, A Simulated Environment Experiment on Annoyance Due to Combined Road Traffic and Industrial Noises, doi:10.3390/ijerph120708413.

CHAPTER 10

Crash Prediction Model for Two-Lane Rural Highways in the Ashanti Region of Ghana

Williams Ackaah[a], Mohammed Salifu[b]

[a] CSIR – Building and Road Research Institute, Kumasi, Ghana

[b] Kwame Nkrumah University of Science and Technology, Kumasi, Ghana

ABSTRACT

Crash Prediction Models (CPMs) have been used elsewhere as a useful tool by road Engineers and Planners. There is however no study on the prediction of road traffic crashes on rural highways in Ghana. The main objective of the study was to develop a prediction model for road traffic crashes occurring on the rural sections of the highways in the Ashanti Region of Ghana. The model was developed for all injury crashes occurring on selected rural highways in the Region over the three (3) year period 2005–2007. Data was collected from 76 rural highway sections and each section varied between 0.8 km and 6.7 km. Data collected for each section comprised injury crash data, traffic flow and speed data, and roadway characteristics and road geometry data. The Generalised Linear Model (GLM) with Negative Binomial (NB) error structure was used to estimate the model parameters. Two types of models, the 'core' model which included key exposure variables only and the 'full' model which included a wider range of variables were developed. The results show that traffic flow, highway segment length, junction density, terrain type and presence of a village settlement within road segments were found to be statistically significant explanatory variables ($p < 0.05$) for crash involvement. Adding one junction to a 1 km section of road segment was found to increase injury crashes by 32.0% and sections which had a village settlement within them were found to increase injury crashes by 60.3% compared with segments with no settlements. The model explained 61.2% of the systematic variation in the data. Road and Traffic Engineers and Planners can apply the crash prediction model as a tool in safety improvement works and in the design of safer roads. It is recommended that to improve safety, highways should be designed to by-pass village settlements and that the number of junctions on a highway should be limited to carefully designed ones.

INTRODUCTION

Road Traffic Crashes (RTCs) continue to be a major socio-economic problem for most developing countries [1] and [2]. In Ghana, it has been estimated that, road traffic crashes cost the nation 1.6% of its Gross Domestic Product [3]. Research on road traffic crashes have shown that crash severity tends to be mostly higher on rural highways than on urban roads [4], [5] and [6]. During a three year period 2005–2007, a total of 21,709 injury road traffic crashes occurred in Ghana resulting in 48,605 casualties and yet these figures may have been higher if under-recording and non-reporting which have been established to be considerable [7] were accounted for in the data. The statistics further indicated that crashes in the non-urban environment (mostly on rural highways) accounted for about 70% of all road traffic fatalities. It is also evident from the data that whilst fatalities on roads in urban settlement areas follow a gradual upward trend, that for the non-urban areas follow a steep trend [4].

Crash Prediction Models (CPMs) have been used elsewhere as a useful tool by road Engineers and Planners. Fletcher et al. [8] found that due to wide differences in traffic mix, road quality, design and road user behaviour, it would be neither valid nor useful to apply simple multiplicative factors or even devise more complex conversion formulae for models developed elsewhere for another country. In Ghana, Salifu [9] developed a crash prediction models for unsignalised urban junctions. Afukaar and Debrah [10] also developed models for predicting crashes for signalised urban junctions. However, there is no study on predicting crashes on rural highways in Ghana.

The study was carried out in the Ashanti Region, the most populous and third largest in terms of land area (24,390 km [2]) of the ten administrative Regions of Ghana. The population of the Region is projected to be about 5,000,000 inhabitants [11] and lies approximately at the centre of the country. The Ashanti Region is one of the Regions with above average Public Health Risk (PHR) of road traffic crashes with a fatality index of 8.8 fatalities/100,000 population per year as against a national average of 8.5 fatalities/100,000 population per year [12].

The objective of the research work was to develop a crash prediction model to identify contributory factors that are likely to affect injury crashes on the rural highways in the Ashanti Region of Ghana. The model was developed for all crashes resulting in death or injury over a three (3) year period 2005–2007 on selected rural highways in the Ashanti Region of Ghana. Damage only crashes were not included in the study because of the high level of under-reporting associated with this type of crashes in Ghana [7]. In this report, a rural highway is defined as a highway passing through the non-urban environment i.e. passing through the village (population of not more than 5000 inhabitants) and rural non-settlement areas.

LITERATURE REVIEW

Substantial research has been conducted over the years on the development of Crash Prediction Models (CPMs) for highway facilities. From the extensive review of literature, the elements considered in this chapter have been identified to be necessary for any model development.

Model form

Miauo and Lum [13] investigated the statistical properties of two conventional linear regression models and identified potential limitations of these models in developing vehicle crashes and highway geometric design relationships. It was demonstrated that the conventional linear regression models lack the distributional property to describe adequately random, discrete, non-negative, and typically sporadic, vehicle crash events on the road. Several other literatures have supported the unsatisfactory property of ordinary linear regression models in developing vehicle crashes, traffic flow and highway geometric design relationships [14], [15] and [16].

Currently, Generalised Linear Regression Model (GLM) is used almost exclusively for the development of Crash Prediction Models [17], [18], [19] and [20]. Sawalha and Sayed[17] said the mathematical form used for any CPM in the framework of GLM should satisfy the following conditions: yield logical results and there must exist a known link function that can linearise the model for the purpose of coefficient estimation.

Error structure

The occurrences of vehicle crashes are discrete random events. That is, the number of vehicles involved in crashes on a given road section during a period of time is probabilistic in nature. Researchers in this field say the

GLM assumes an error structure (probabilistic structure) that is used in the data generation process (crashes). This distributional assumption is also used to obtain tests and confidence statements about the estimated regression coefficients.

According to Lord et al. [21] a crash is in theory, the result of a Bernoulli trial. From literature, the appropriate probability model that accounts for a series of Bernoulli trials is known as the binomial distribution [21] and [22]. For typical motor vehicle crashes which are rare and therefore the event has a very low probability of occurrence and a large number of trials exists, the binomial distribution is approximated by a Poison distribution. However, one limitation of the Poisson distribution is that the mean is assumed to be equal to its variance. Many previous studies have found that crash data tend to be over-dispersed in many situations with the variance being significantly higher than the mean. In such cases, any inferences made based on Poisson model estimation may lead to wrong conclusions. As a result of this, many researchers recommend using alternative methods such as Negative Binomial distribution which does not require the equal mean and variance assumption [13], [14], [20] and [21].

Model goodness of fit assessment

Researchers in this area have identified two major statistical measures used in assessing the goodness of fit in both Poisson and Negative Binomial approaches. These are Pearson Chi-square statistic and Deviance.

To measure the overall goodness-of-fit in linear regression models, the coefficient of determination, R-squared is often used. The ordinary R-squared cannot be used for Poisson and Negative Binomial models because R-squared is maximised by ordinary least squares whiles in the Poisson and Negative Binomial models, the Maximum Likelihood estimation method is usually used. FridstrØm et al. [23] developed several alternative goodness-of-fit methodologies for generalised Poisson regression models. Miaou [24] also investigated different approaches to calculate R-squared values for different regression techniques using different distribution assumptions including Poisson and Negative Binomial.

Outlier analysis

Data collected may contain odd or extreme observations called outliers either because errors took place during data collection and recording or they are really different from the rest of the data [17]. These extreme observations may have effect on the model equation if not managed well.

Outlier analysis could be carried out to identify these influential observations. However, Sawalha and Sayed [17] emphasised that excluding influential outliers from the development of a crash model is not synonymous with neglecting these outliers or removing them from the database. Rather, they should be investigated to determine what makes them different and whether any information can be extracted from them.

Selecting the model explanatory variables

According to Sawalha and Sayed [17] model generality requires that a model be developed in accordance with the principle of parsimony, which calls for explaining as much of the variability of the data using the least number of explanatory variables. Deciding on which variables to include in a model is a multi-step process. First, variables believed to relate to the phenomenon in question have to be determined. This is usually based on previous studies, experience and engineering judgement based on available information [17] and [25]. The manner in which these variables will be used in the model must also be determined. Last is the statistical determination as to which variables should remain in the model. Many studies have suggested that the t-ratio of its estimated parameter should be significant at the 95% confidence level, i.e. p-values < 0.05[17],[19], [20] and [25].

Delineation of road sections

Miaou and Lum [13] considered two (2) methods, i.e. fixed-length sections, or homogeneous sections. They found out that, short road sections had undesirable impacts on the estimation of their linear regression models. Longer sections were, therefore, suggested to be preferred. However, given that most of the curved and graded sections were relatively short, the analysts oftentimes were unable to find a sufficient number of long and homogeneous road sections, which exhibited wide enough variations in geometric design variables to study the relationships. To overcome this problem, many analysts according to Miaou and Lum [13] have chosen to keep long road sections and not to insist on having homogenous road sections.

Resende and Benekohal [26] reached the conclusion that, to get a reliable crash prediction models crash rates should be computed from 0.8 km or longer sections.

METHODS FOR DATA COLLECTION

Range of data collected

Variables believed to relate to crashes on rural highways were identified. This was primarily based on literature and engineering judgement based on available information and exploratory analysis. The data collected for all sections for the study included: injury crash data (for a 3 year period, 2005–2007), traffic flow data, vehicle speed data, road characteristics and road geometry. The study was approved by the Institutional Review Board (IRB) of the Kwame Nkrumah University of Science and Technology, Kumasi, Ghana.

Selection of road corridors

An extensive reconnaissance survey was carried out to identify suitable sites from which the sample of sites for the study was drawn. The road sections from which the samples were drawn were all two-lane single carriageway rural segments of road. Stratified random sampling technique was used in the road selection process in order to select segments from the population to include roadways from the three (3) different highway categories namely: national roads, N (which link regional capitals to the national capital); inter-regional roads, IR (which link regional capitals to one another); and regional roads, R (which link the district capitals to the regional capitals) so as to avoid bias.

Only road sections which had not undergone any modification during the period (2005–2007) for which crash data were studied were included in the sample. Since the objective of the study was to establish the effects of the different variables on crashes, selection of segments were done in order to achieve variety in traffic volumes, speeds, road characteristics and geometry.

Selection of road links

A total of fourteen (14) road corridors were selected for the project. These road corridors were subdivided into seventy-six (76) links and studied for the purpose of model development. Links were either taken as between two major junctions, between two towns, between a junction and a town or between a junction or town and any major landmark, for example, a river/bridge.

Thirty-one of the sites representing 40.8% were on National Roads, whilst 21(27.6%) were on Inter-Regional Roads and 24 (31.6%) were on Regional Roads. The total road length considered was 186.3 km long with

National Roads, Inter-Regional Roads and Regional Roads constituting 41.3%, 31.4% and 27.3% respectively.

Road traffic crash and traffic flow data

In Ghana, information on road traffic crashes is available at the Building and Road Research Institute (BRRI). The database is compiled from police files using a standard crash report form. This form includes information about the crash location, the vehicle(s) and casualties involved. Most of the information has been coded and stored in computers at the BRRI using the Micro-computer Accident Analysis Package (MAAP) software. For the purpose of this study, the relevant information for each crash was the location and date. Crash data were retrieved and analysed with the help of the kilometre analysis facility available in the MAAP software. A total of 301 crashes were recorded over the period 2005–2007 on road links considered in the study of which 50.8% occurred on roads classified as National Highways, 30.2% occurred on the Inter-Regional Highways and 18.9% on the Regional Highways.

Traffic flow data including Average Daily Traffic (ADT) was obtained from the Traffic and Planning Unit of the Ghana Highway Authority (GHA) for the different road corridors for the study. Traffic volume for all the road categories considered ranged from 347 to 8948 vehicles per day.

Procedure for speed measurement

Speeds measurements were made with a radar gun at not less than three (3) locations within each segment in order to get a representative speed for the segments which oftentimes were un-homogenous. The data collection team was stationed in a parked vehicle and used the speed gun to measure spot speeds of vehicles. Effort was made to measure the speeds of all passing vehicles. In a situation where there was a platoon of vehicles, only the lead vehicle in the fleet was measured since vehicles following in a platoon do not travel at their desired speeds. Mean speeds of 83.5 km/h, 81.3 km/h and 69.2 km/h were recorded on National, Inter-Regional and Regional highways respectively.

Road characteristics and geometric data

Field surveys were carried out by a trained data collection team led by the Project Leader for all selected road segments for the purpose of establishing the road characteristics. Measurement of link length was done on site. Carriageway and shoulder widths were also measured on site and at more than two (2) locations within a link in order to get a representative figure for the entire segment. The variables presented in Table 3.1 were collected for the study.

Table 3.1. Road characteristics and geometric data collected for the study.

Variable description	Range of values
Categorical data	
Terrain type	0 = Hilly/Rolling; 1 = Flat
Road marking	0 = Absent; 1 = Present
Shoulder	0 = Absent; 1 = Present
Should type	0 = Unpaved; 1 = Paved
Village settlement within road segment	0 = Absent; 1 = Present
Road surface type	0 = Surface dressing; 1 = Asphalt
Continuous data	
Length of section	0.8 km–6.7 km
Carriageway width	7.0 m–7.8 m
Shoulder width	1.0 m–2.5 m
Junction density within segment	0–1.7 per km
Access density within segment	0–17.5 per km
Horizontal curves density within segment	0–6.4 per km
Vertical curves density within segment	0–3.6 per km

Accesses were defined as any road connecting a private property to a public road. Junctions consisted of T-junctions, Y-junctions and X-junctions within segments.

MODEL DEVELOPMENT

Multiple regression analysis

The main objective of modelling with several variables simultaneously was to permit greater insight into the relative effects of the different highway variables on crashes. Modelling was undertaken at two stages — the 'core' model and the 'full' model all within the framework of Generalised Linear Models (GLMs). The 'core' model included exposure variables only which in this case were the traffic flow and section length. Apart from the fact that 'core' models have the advantage of being simple in form, they are also useful as a rough guide for identification of locations with a high frequency of crashes (blackspots), as well as for the prediction of the effect of traffic flow changes on crash occurrence [9].

It is also important to know that there are a large number of variables apart from traffic flow and length that might contribute to crashes. These variables will not be in the list of factors included in the 'core' model. The 'full' model which was developed included a wider range of other important causal variables.

Following the knowledge established from the review of previous work the general form of the 'core' and 'full' models developed under this study were:

'Core Model' — exposure variables only.

$$E(Y) = a_o L^{a1} Q^{a2} \tag{4.1}$$

'Full Model' — product of the powers of the exposure variables multiplied by an exponential function incorporating the remaining explanatory variables.

$$E(Y) = a_o L^{a1} Q^{a2} \exp \textstyle\sum_j b_j x_j \tag{4.2}$$

where:

$E(Y)$=predicted crash frequency,

L=section length (km),

Q=ADT (per day),

xj=is any variable additional to L and Q, and

exp=exponential function, e = 2.7183

a_0, a_1, a_2, $bj$are the model parameters.

In accordance with the GLM framework, Eqs. (4.1) and (4.2) are transformed into the prediction mode using a log-link function as follows:

'Core Model'

$$\ln[E(Y)] = \ln(a_o) + a_1\ln(L) + a_2\ln(Q) \tag{4.3}$$

'Full Model'

$$\ln[E(Y)] = \ln(a_o) + a_1\ln(L) + a_2\ln(Q) + \sum_j b_j x_j \tag{4.4}$$

Modelling procedure

The mean and variance for the crash data were 3.96 and 8.28 respectively which indicated that the data set was over-dispersed since the variance is greater than the mean. Initial modelling using Poisson error structure also showed that the estimated dispersion parameter (*Φ*) defined as:

$$\Phi = \frac{Pearson\ \chi^2}{(N-p)} \quad \ldots \quad \ldots \quad \ldots \quad \ldots \quad \ldots \tag{4.5}$$

where N is the total number of sections and p is the number of parameters in the model was greater than one (1) indicating that the data set was over-dispersed [27]. That means Poisson distribution is not capable of explaining the true distribution underlying the crash frequency.

Generalised Linear Model (GLM) was used to estimate the model coefficients using the STATA software package and assuming a Negative Binomial error distribution, all consistent with earlier research works in developing these models. By specifying the dependent variable, the explanatory variables, the error structure (in this case the Negative Binomial) and the link function (in this case log), the model is fitted. Model parameters (coefficients) were estimated using maximum likelihood approach. The procedure which was adopted in the model development was the forward procedure in which the variables were added to the model one by one.

Model evaluation

In consonance with earlier studies, the decision on which variables should be retained in the model was based on two criteria. The first criterion was whether the t-ratio of the estimated parameter was significant at the 95% confidence level (p-value less than 5%). The second criterion was whether the addition of the variable to the model cause a significant drop in the scaled deviance at the 95% confidence level.

Two statistical measures were used in assessing the validity of the model developed. These were the Pearson Chi-square statistic and Deviance statistic. The coefficient of determination (R^2) was also employed to determine the amount of variability in the response variable explained by the variation in the selected set of explanatory variables. The R-squared estimation was carried out by the method recommended by Miaou [24].

Variables included in the model

Although a large number of variables were collected and considered for inclusion in the 'full' model development, only variables with significant estimated parameter coefficients (p-values less than 5%) were maintained in the model. This method is similar to the one used by Vogt and Bared [28]. The variables shown in Table 4.1 were thus, used in the full model development.

Table 4.1. Description of variables and symbols used for crash prediction models.

No.	Variable name	Description	Variable type	Symbol
1	lnvolume	The logarithm of average daily traffic	Continuous	Q
2	lnlengthkm	The logarithm of link length (km)	Continuous	L
4	juncdensity	Junction density	Continuous	JDen
5	terraintype	Terrain type	Categorical (1 — flat, 0 — otherwise)	TerTyp
6	presenc~ment	Presence of village settlement	Categorical (1 — present, 0 — absent)	VSet

MODEL RESULTS AND INTERPRETATION

'Core' model

Parameter estimations for the log-linear equation for the 'core' model using Negative Binomial error structure are as shown in Table 5.1.

Table 5.1 Parameter estimation for 'core' model as estimated by STATA statistical software.

				Log likelihood = − 163.89315		
Variable	Coefficient	Standard error	Chi-square statistic, Z	P > |Z|	95% conf. limits	
					Lower	Upper
lnvolume	0.37107	0.09064	4.09	0	0.1934	0.5487
lnlength	0.36661	0.14182	2.59	0.01	0.0887	0.6446
_cons	− 1.92725	0.7321	− 2.63	0.008	− 3.36 21	− 0.49 24

Goodness of fit measures: deviance (value/d.f.) = 77.7200 (1.0647); Pearson chi-square (value/d.f.) = 79.62985 (1.09082).

The resulting 'core' model has been determined to be as follows:

$$E(Y)=0.1\times L^{0.3666}\times Q^{0.3711}$$

where:

E(Y)=expected crashes along the road segment for 3 years,

L=length (km) of road segment and

Q=Average Daily Traffic (ADT)

The goodness-of-fit statistics for the model shows that the model fits reasonably well with the data. The Pearson Chi-square and Deviance statistics divided by its degrees of freedom were estimated to be 1.09 and 1.06 respectively as shown in Table 5.1. The values are within the permissible range [25] of 0.8 and 1.2 indicating that the Negative Binomial distribution assumption is acceptable.

The Average Daily Traffic (ADT) and segment length were statistically significant ($p < 0.05$) with positive estimated model parameters in the 'core' model. This indicates that the crash frequency increases with an increase in the traffic flow or segment length whilst the other variables are held constant. The exponent on segment length was 0.37. Since distance travelled is a measure of exposure, it follows that the more distance travelled, the more likely one is involved in a crash. The exponent on traffic flow (ADT) was also estimated to be 0.37. This figure seems to be small when compared with other studies elsewhere [25], [28] and [29] and considering the importance of traffic flow as a major determinant of road traffic crashes. Qin et al. [29] when they analysed different crash types found the lowest exponent of 0.4 on Average Annual Daily Traffic (AADT) when they considered only single-vehicle crashes. In the crash statistics on rural highway segments sampled for the study, over two-thirds (68.1%) involved single-vehicles and this may partly explain the rather low exponent value for the ADT.

Using the R-squared value, the variables in the flow based model ('core' model), namely traffic flow and segment length could only explain 31.3% of the total variation in the crash data. According to Salifu [9], 'core' models may be regarded as relatively coarse and rough estimators of crash frequency.

'Full' model

Parameter estimations for the log-linear equation for the 'full' model using negative binomial error structure are as shown in Table 5.2.

Table 5.2. Parameter estimation for 'full' model as estimated by STATA statistical software.

Variable	Coefficient	Standard error	Chi-square statistic, Z	log likelihood = − 143.43250		
				P > |Z|	95% conf. limits	
					Lower	Upper
lnvolume	0.24793	0.813	3.05	0.002	0.0886	0.4073
lnlengthkm	0.50599	0.13304	3.8	0	0.2452	0.7667
juncdensity	0.27786	0.14124	1.97	0.049	0.001	0.5547
terraintype	0.45512	0.12866	3.54	0	0.203	0.7073
presenc~ment	0.47174	0.13591	3.47	0.001	0.2054	0.7381
_cons	− 1.68896	0.6416	− 2.63	0.008	− 2.9465	− 0.4315

Goodness of fit measures: deviance (value/d.f.) = 56.8476 (0.8121); Pearson chi-square (value/d.f.) = 56.67047 (0.80958).

The resulting 'full' model has been determined to be as follows:

$$E(Y) = 0.1847 \times L^{0.5060} \times Q^{0.2479} \times EXP^{(0.2779\ JDen + 0.4551\ TerTyp + 0.4717\ VSet)}$$

where:

E(Y) expected crashes along the road segment for 3 years,

L length (km) of road segment and

Q Average Daily Traffic (ADT)

JDen Junction Density

TerTyp Terrain Type (1 — flat, 0 — otherwise)

Vset Presence of Village Settlement (1 — present, 0 — absent)

EXP Exponential function, e = 2.718282.

The goodness-of-fit statistics for the full model shows that, the model fits reasonably well with the data. The Pearson Chi-square and Deviance statistics divided by its degrees of freedom were estimated to be 0.81 as shown in Table 5.2. In other words, the estimated values of the Pearson Chi-square and Deviance divided by the degrees of freedom were within the permissible range (i.e. between 0.8 and 1.2) indicating that the Negative Binomial distribution assumption is acceptable.

The R-squared value for the full model seems to be reasonable as it could explain 61.2% of the variation in the crash data. Dissanayake and Ratnayake [25] have said the coefficient of determination is significant if found to be greater than 0.45. Approximately 39% of the systematic variations in the data set remain unexplained by the model developed. The 'percentage unexplained' may be attributed to human behaviour which has increasingly been acknowledged as one of the predominant factors in road traffic crashes and factors relating to the vehicle.

Significant explanatory variables

The following sections consider the significant variables identified through the 'full' model results.

Traffic flow and safety

The Average Daily Traffic (ADT) in the 'full' model was also significant with positive estimated model parameter. From the model, an increase in the traffic flow by 50% is expected to cause an increase in road traffic crashes by 11% whilst doubling the number of vehicles on the road is expected to cause an increment of 19%. Meanwhile, the proportions of heavy goods vehicles were found not to have a significant effect on crashes in this study.

Road segment and safety

The length of road segment, the other important exposure variable aside traffic volume was the most significant predictor in the 'full' model with a positive estimated exponent of 0.51. This was as expected as distance travelled is a measure of exposure therefore it follows that the longer the distance travelled, the more likely one is involved in a crash.

Junction density and safety

The number of junctions per unit length within a road segment has a considerable influence on the crash risk as indicated by the estimated model parameter. In general, the more the number of junctions per unit length of segment, the higher the crash risk. For example, for all other variables held constant, adding one junction to 1 km section of road segment increases crashes by 32.0%. The number of junctions on a highway should therefore be limited to carefully designed ones to improve road traffic safety in Ghana.

Terrain type and safety

The analysis suggests that, flat terrains record more injury crashes when compared with rolling or hilly terrains. An increase of 57.6% in injury crashes occurred at links with flat terrain compared with links with rolling or hilly terrain. This probably had to do in part, with the fact that drivers tend to travel at higher speeds on straight and flat road sections than on rolling or hilly terrains.

Presence of village settlement and safety

The explanatory variable for presence of village settlement proved to be very important in the model. An increase of 60.3% in injury crashes were found for sections which had a village settlement within them compared with segments with no settlements.

From the crash data used in the model development, 90 (29.9%) were 'hit pedestrian' collisions of which 81.1% occurred in the settlement areas. It was presumed that the variable 'presence of village settlement' represents the level of pedestrian interference with vehicles on the highways. The interaction between fast moving vehicles and pedestrians is undesirable and contribute to the high number of injury crashes in the settlement areas.

More fatal pedestrian collisions have been reported in the village settlements along the major highways in Ghana [30].

Non-significant explanatory variables

The following sections also look at the variables which were considered but were not significant in the 'full' model results.

Horizontal and vertical curves

Horizontal curvature has been found to influence road traffic crashes in other studies[31] and [32]. Sharper and longer curves results in more crashes because of the additional effort required by the driver to control the stability of the vehicle. Again, studies have shown that reducing gradients reduces the number of crashes [31] and [32]. The absolute number of curves per segment length used in this study did not capture the effect of the individual degrees of curvatures of the curves. Construction plans (horizontal and vertical profiles) for most of the roads selected for the study were not readily available for the curvatures to be extracted from them. Attempt to use the tract function of a Global Positioning System (GPS) device to measure tracks along the road segments from which the vertical and horizontal curvatures could have been calculated proved futile. It was found that the data were not of sufficient completeness or accuracy to be included in the analysis as in particular, the vertical curvature of relatively flat sections was often lost. Fletcher et al. [8] also experienced a similar difficulty. This may explain why both vertical and horizontal curvatures did not show to be statistically significant in the study.

Speed

Recent studies have concluded that crash involvement increases with speed on any particular road or for any particular driver [18], [33] and [34]. Mean speed was not a useful explanatory variable in this study as it correlated highly with the variables presence of village settlement and terrain type. Standard deviation speed and 85th percentile speed were also tested in the model but were not statistically significant.

Presence of shoulder, shoulder type, shoulder width and road width

The effect for the variable presence of shoulder on crash rates could not be assessed as there was no variability in the data collected as the entire road links sampled had shoulders. Shoulder type has also been found in other studies to be important in terms of stability of vehicles when a vehicle goes off the road. However, the variable shoulder type did not prove to be a useful explanatory variable as it correlated with traffic flow. Adequate shoulder width has also been established to be necessary for safe recovery when vehicles veer off the road [32]. The variable shoulder width was not statistically significant when data from all segments (which included paved and unpaved shoulders) was used in the model. This may be explained by the fact that paved and unpaved shoulders have different effect on crashes. Two-thirds (n = 51) of the segments had paved shoulders while one-third (n = 25) had unpaved shoulders. Efforts to develop models for the sub-groups (paved and unpaved) also did not yield any significant results as the range of paved shoulder width encountered was limited (2.0 m, 2.2 m or 2.5 m) and the sample size for unpaved shoulders was small.

Generally it has been found that crash rates decrease when lane width increases[28] and [32]. However, road width was not found to be a useful explanatory variable in this study as it also correlated highly with traffic flow.

Road marking
Marked roads have been found to be safer than unmarked roads [8]. Most of the road segments sampled were delineated (n = 62, 82%) while only a few (n = 14, 18%) had no marking. The comparative small sample size may have resulted in this variable not being a significant explanatory variable in this study.

CONCLUSIONS AND RECOMMENDATIONS

Conclusions

Statistical models have been developed to predict crashes on rural highways in the Ashanti Region of Ghana. Generalised Linear Model (GLM) with Negative Binomial (NB) error structure was used to estimate the model parameters. Two types of models, the 'core' model which included key traffic exposure variables only and the 'full' model which included a wider range of variables were developed.

From the model developed, traffic flow, segment length, junction density, terrain type and presence of village settlement within road link are significant explanatory variables that influence the prediction of injury crashes on the rural highways in the Ashanti Region of Ghana. The study proved that, passing a highway through settlement areas, traffic volume, provision of long straight and flat sections and increasing number of junctions per unit road length tend to worsen road traffic safety.

Road and Traffic Engineers and Planners can apply the crash prediction model as a tool in safety improvement works and in the design of new safer roads.

Recommendation

From the model developed and subsequent discussions above, the following recommendations are drawn:

- Safety situations on the highways should be improved through the provision of by-passes around the village settlements.
- The number of junctions on a highway should be limited to carefully designed ones to improve road traffic safety in Ghana.
- Road Engineers should as much as possible avoid provision of long straight and flat sections on highways to improve traffic safety. Intermittent graceful curves could be introduced to break the monotony.

ACKNOWLEDGEMENTS

The authors would like to acknowledge the Road Traffic Injury Research Network (RTIRN) for partly sponsoring this research work through their scholarship programme — "Grants for Junior Researchers" with funding support from the World Bank Global Road Safety Facility. The opinions, findings and conclusions expressed here are those of the authors.

REFERENCES

1. G.D. Jacobs, A. Aeron-Thomas, A. Astrop, Estimating global road fatalities, TRL Report 445, Transport Research Laboratory, Crowthorne, Berkshire, 2000.
2. C. Ghee, D. Silcock, A. Astrop, G. Jacobs, Socio-economic aspects of road accidents in developing countries, TRL Report 247, Transport Research Laboratory, Crowthorne, Berkshire, 1997.
3. F.K. Afukaar, W. Agyemang, E.K. Debrah, W. Ackaah, Estimation of the cost of road traffic accidents in Ghana, Ministry of Transportation, National Accident Management Project, Final Report, 2006.

4. F.K. Afukaar, W. Agyemang, W. Ackaah, I. Mosi, Accident Statistics 2007. Building and Road Research Institute, Council for Scientific and Industrial Research, Kumasi, 2008.
5. K. Hamilton, J. Kennedy, Rural road safety: a literature review, Scottish Executive Social Research (2005).
6. G.D. Jacobs, A study of accident rates on rural roads in developing countries, TRRL Report, Transport and Road Research Laboratory, Crowthorne, Berkshire, 1976, p. 732.
7. F.K. Afukaar, W. Ackaah, W. Agyemang, Estimation of levels of under-reporting of road traffic accidents in Ghana, Ministry of Transportation, National Accident Management Project, Final Report, 2006.
8. J.P. Fletcher, C.J. Baguley, B. Sexton, S. Done, Road accident modelling for highways development and management in developing countries, Main Report: Trials in India and Tanzania, Project Report No: PPR095, , 2006.
9. M. Salifu, Accident prediction models for unsignalised urban junctions in Ghana, International Association of Traffic and Safety Science (IATSS) Research 28 (1) (2004) 2004.
10. F.K. Afukaar, E.K. Debrah, Accident prediction models for signalized intersections in Ghana, Ministry of Transportation, National Accident Management Project, Final Report, 2007.
11. Ghana Statistical Service, 2000 population and housing census, Special report on 20 largest localities, 2002.
12. M. Salifu, I. Mosi, K. Addae-Bofah, K. Sarpong, Accident Statistics 2005. Building and Road Research Institute, Council for Scientific and Industrial Research, Kumasi, 2006.
13. S.-P. Miaou, H. Lum, Modelling vehicle accidents and highway geometric design relationships, Accident Analysis and Prevention. 25 (6) (1993) 689–703.
1. 1 4. S.-P. Miaou, The relationship between truck accidents and geometric design of road sections: Poisson versus negative binomial regressions, Accident Analysis and Prevention. 26 (4) (1994) 471–482.
14. S.H. Kim, S.B. Chung, K.H. Song, K.S. Chong, Developing of an accident prediction model using GLIM (generalized log-linear model) and EB method: a case of Seoul, Journal of the Eastern Asia Society for Transportation Studies 6 (2005) 3669–3682.
15. N.J. Garber, L. Wu, Stochastic models relating crash probabilities with geometric and corresponding traffic characteristics data, Research Report No. UVACTS-5-15- 74, 2001.
16. Z. Sawalha, T. Sayed, Traffic accident modeling: some statistical issues, Canadian Journal of Civil Engineering 33 (2006) 1115–1124.

17. M.C. Taylor, A. Baruya, J.V. Kennedy, The relationship between speed and accidents on rural single-carriageway roads, TRL Report TRL511, Transport Research Laboratory, Crowthorne, Berkshire, 2002.
18. S. Harnen, R.S. Radin Umar, S.V. Wong, W.I. Wan Hashim, Development of prediction models for motorcycle crashes at signalized intersections on urban roads in Malaysia, Journal of Transportation and Statistics V7 (2004) N2/3.
19. E.T. Donnell, J.M. Mason, Predicting the frequency of median barrier crashes on Pennsylvania interstate highways, Accident Analysis and Prevention, 38, 2006, pp. 590–599.
20. D. Lord, S.P. Washington, J.N. Ivan, Poisson, Poisson-gamma and zero-inflated regression models of motor vehicle crashes: balancing statistical fit and theory, Paper Accident Analysis and Prevention, 2004, p. 03225.
21. J.T. McClave, T. Sincich, Statistics, Tenth Edition. Pearson Education, Inc0-13- 200302-3, 2003.
22. L. FridstrØm, J. Ifver, S. Ingebrigsten, R. Kulmala, L.K. Thomsen, Measuring the contribution of randomness, exposure, weather, and daylight to the variation in the road accident counts, Accident Analysis and Prevention 27 (1) (1995) 1–20.
23. S.-P. Miaou, Measuring the goodness-of-fit of accident prediction models, FHWA-RD-96-040, Federal Highway Administration, Washington, D.C, 1996.
24. S. Dissanayeke, I. Ratnayake, Statistical modelling of crash frequency on rural freeways and two-lane highways using negative binomial distribution, Advance in Transportation Studies an International Journal (2006)8 Section B 9.
25. P. Resende, R. Benekohal, Effects of roadway section length on accident model, Traffic Congestion and Traffic Safety in the 21st Century Conference, American Society of Civil Engineers (ASCE), Chicago, IL, 1997.
26. P. McCullar, J.A. Nelder, Generalized Linear Models, Chapman and Hall, London, 1989.
27. A. Vogt, J.G. Bared, Accident models for two-lane rural roads: segments and intersections, Publication No. FHWA-RD-98-133, 1998.
28. X. Qin, J.N. Ivan, N. Ravishanker, Selecting exposure measures in crash rate prediction for two-lane highway segments, Accident Analysis and Prevention 36 (2004) 183–191.
29. W. Ackaah, D.O. Adonteng, Analysis of fatal road traffic crashes in Ghana8 First published on 20 may 2010, International Journal of Injury Control and Safety Promotion (2010) 1–78 iFirst.
30. R. Elvik, T. Vaa, The Handbook of Road Traffic Safety Measures, Elsevier Ltd978- 0-08-044091-0, 2004.

31. S.-P. Miaou, P.S. Hu, T. Wright, S.C. Davis, A.K. Rathi, Development of relationship between truck accidents and geometric design: phase I, Publication No. FHWA-RD-91-124, 1993.
32. C.N. Kloeden, G. Ponte, A.J. McLean, Travelling speed and the risk of crash involvement on rural roads, Road Accident Research Unit, Adelaide University, 2001.
33. A. Quimby, G. Maycock, C. Palmer, S. Buttress, The factor that influence a driver's choice of speed — a questionnaire study, TRL Report 325, Transport Research Laboratory, Crowthorne, Berkshire, 1999.

CITATION

Williams Ackaah, Mohammed Salifu, Crash prediction model for two-lane rural highways in the Ashanti region of Ghana, IATSS Research, Volume 35, Issue 1, July 2011, Pages 34-40, ISSN 0386-1112, http://dx.doi.org/10.1016/j.iatssr.2011.02.001.

Index

S

T

U

V

W